Marion Ettlinger

About the Author

ELINOR BURKETT, a native of Philadelphia, has worked as a newspaper reporter, university professor, and magazine writer. She and her husband took their first vacation together in Albania, honeymooned in Havana, and count among their fondest memories the week they eluded the Mauritanian secret police. Once longtime residents of the Catskill Mountains of New York, they now make their home in Fairbanks, Alaska, where she is chairing the Department of Journalism at the University of Alaska.

ALSO BY ELINOR BURKETT

Another Planet:
A Year in the Life of a Suburban High School

The Baby Boon: How Family-Friendly America
Cheats the Childless

The Right Women: A Journey Through the
Heart of Conservative America

Consumer Terrorism (with Frank Bruni)

The Gravest Show on Earth: America in the Age of AIDS

A Gospel of Shame: Children, Sexual Abuse,
and the Catholic Church (with Frank Bruni)

So Many Enemies, So Little Time

AN AMERICAN WOMAN
IN ALL THE WRONG PLACES

Elinor Burkett

Perennial

An Imprint of HarperCollins*Publishers*

For Dennis,

who carries all my baggage

Sections of this book appeared in slightly different form in *Elle* and *Penthouse*.

A hardcover edition of this book was published in 2004 by HarperCollins Publishers.

HarperCollins books may be purchased for educational, business, or sales promotional use. For information please write: Special Markets Department, HarperCollins Publishers Inc., 10 East 53rd Street, New York, NY 10022.

FIRST PERENNIAL EDITION PUBLISHED 2005.

DESIGNED BY SARAH MAYA GUBKIN

Library of Congress Cataloging-in-Publication Data is available upon request.

ISBN 0-06-052443-X (pbk.)

05 06 07 08 09 ❖/RRD 10 9 8 7 6 5 4 3 2 1

Contents

So Many Enemies,
So Little Time

PROLOGUE

The church is close, but the road is icy.
The bar is far, but I will walk carefully.
———RUSSIAN PROVERB

Just after 7 P.M. on September 18, 2001, U.S. ambassador John O'Keefe strode to the podium in the overflowing Shayr Ballroom of the Hyatt Regency Hotel in Bishkek, Kyrgyzstan, an outpost of first-world luxury in a Central Asian capital disintegrating in post-Communist neglect.

Slight and soft-spoken, O'Keefe was an old hand at dicey situations, having charted the safety of the American community through the chaos of the breakup of the former Yugoslavia, then through the evacuation of Americans from Tajikistan. He would need all that skill to steer his anxious audience of 400 between calm and a full acceptance of the gravity of the situation they were facing.

"Right now, we know of no specific threat to U.S. citizens or facilities in this country," he said after leading in a moment of silence to honor our dead. "The situation is more dangerous than it was before, but we're not telling you to leave."

The task of explaining how to cope with that greater danger fell to Paul Avallone, the embassy's regional security officer, and he was typically clipped and concise. Number one: Vary your routine. Number two: Watch your back. Number three: Do not, under any circumstance, gather in groups or in places that are known American hangouts. At that, a few nervous titters broke the silence, although Avallone seemed oblivious to the irony of delivering that admonition to a large group of Americans called together for an emergency town hall meeting at the most American venue in town.

The crowd turned sober as denial dissolved.

"But have there been anti-American incidents? What about the library at the American University?"

"Is there any sign that Mujahideen rebels from the Islamic Movement of Uzbekistan (IMU) have crossed the border?"

"What do we know about the relationship between the IMU and the Taliban?"

By then, most of the ubiquitous cellphones had been turned off.

"Our policy is that we practice no double standard in information," O'Keefe reassured the group. "What we know, you will know."

O'Keefe didn't know much, which left each of us free to indulge our own fantasies, cheery or dire. The room sparked with a dozen rumors: The Islamic Movement of Uzbekistan, an Al-Qaeda affiliate which had declared a holy war against Central Asian governments in its quest to Afghanize the region, was massing for an attack. Anti-American leaflets were being distributed by Hizb-ut-Tahrir, a Kyrgyz Muslim political party which had just demanded that the government justify its refusal to recognize the Koran as the only true constitution. In the south, along the Uzbek border, where the hand of Islam was an increasingly heavy social weight, mullahs were calling the faithful to jihad.

Bishkek felt swathed in everyday calm, the mundane rhythms of life unbroken by anything more jarring than buses belching out the music of Third-World streets and milkmen crying, "Moloka, smetana" ("Milk, sour cream"). No mobs chanted "Death to the Infidels." But sitting just 350 miles north of the Afghan border, we

were tensed for the shock wave we knew would be set off the
moment the United States began unleashing its bombs on Osama
bin Laden's protectors.

"Is there any chance we'll have to be evacuated?" someone
yelled from the back of the room. Waiters had just delivered trays of
finger sandwiches and raspberry custard tarts to tables in the corners of the ballroom, but the word *evacuation* dulled the appetite.

"There's always a chance, and we're setting up procedures,"
O'Keefe replied quietly, as if to cool down the room. The content
of his statement had the opposite effect.

"How much notice will we have?"

"How will we be notified?"

"Will you use commercial flights or charters?"

Everyone yelled at once. Manas International Airport wasn't
exactly a hub. British Air flew to London on Sunday, Tuesday and
Friday, and Turkish Air to Istanbul on Wednesday and Saturday.
Aeroflot offered service to Moscow four days a week, but the U.S.
government didn't exactly endorse the safety of Aeroflot. No one
bothered to mention the weekly flights to Urumqi, China, or to
Dushanbe, Tajikistan. That would have been flying out of the frying
pan directly into the field of fire.

O'Keefe was banking on moving people out through Almaty,
Kazakhstan, a 165-mile, three-hour drive to the north. The Almaty
airport offered daily flights to London and Amsterdam, Istanbul,
India, and Germany.

"But most of us don't have Kazakh visas," someone moaned.

Theoretically, the Kazakhs provided transit visas for foreigners
crossing the country in less than seventy-two hours. But in Kazakhstan, the distance between theory and reality was the mood or
the greed of the border guard.

In perfect sync with the rising tide of groans, the vice consul
grabbed the microphone. "Don't worry, I had a long lunch, with a
whopping lunch bill, with the Kazakh vice consul. There will be
no problem with visas for Americans."

Other problems weren't so easily resolved.

"But my husband isn't American, what about him?"

"Yeah, are you going to help spouses who aren't U.S. citizens?"

The small European contingent—British, French and Swiss stranded without a local embassy—sheepishly inched forward on their seats, hoping for some encouraging words of inclusion.

The room sizzled with a dozen conflicting emotions, from skepticism to hysteria, boiling out of the impossible stew of embassy personnel and missionaries so sensitive about preaching the gospel in a Muslim nation that they never uttered what they called "the M-word," of thirty-something consultants training the locals in constitution building and the fundamentals of profit margins, Peace Corps volunteers, nurses, middle-aged men courting Internet brides and Fulbright professors like me. Isolated from all but the other Soviet republics until the Russians lost their financial stamina, if not their appetite, for imperialism, Kyrgyzstan, sandwiched in between China, Kazakhstan and Uzbekistan, had become a frontier, a magnet for latter-day pioneers hawking everything from democracy and Christianity to computers and acne concealer.

Three middle-aged women sitting in the row of chairs in front of me huddled close together. "I hear that the British Air flights are booked solid for the next month," said one, two infants sleeping in her lap. "I don't want to get stranded here."

Her friend draped a comforting arm across her shoulders, and said shakily, "I can handle stranded. It's being kidnapped that terrifies me."

A clutch of Peace Corps volunteers, yanked out of their villages and admonished to remain in the capital indefinitely, rolled their eyes at that exchange.

"This is so stupid," complained a bearded young man dressed in a vaguely Pashtun outfit and embroidered Muslim cap that had ceased to be hip a week earlier. "What's a little danger?"

At the podium, the vice consul ran down his list of evacuation advice. "Keep a bag packed with essentials so that you are ready to move at a moment's notice. And just one bag, that's it. Don't try to take everything with you.

"If we need to get out, your warden will notify you, so make sure he or she has all your phone numbers. And if you can't get to the embassy, stay home no matter what's going on outside! When we can, we'll send marine guards or someone to get you."

Voices rose as reality set in. A woman sitting by the windows slumped over and cried out, mostly to herself, "What if there aren't enough planes?"

O'Keefe regained the mike. "Remember, folks, this is the worst-case scenario. None of this is likely to happen. If things aren't looking good, we'll start drawing down with a voluntary evacuation. Calm down, we'll get through this. Just stay out of the South along the Uzbek and Tajik borders."

I felt my stomach drop. That, as it happened, was exactly where my husband, Dennis, was at that moment.

I panic easily where Dennis is concerned, envisioning him dead by the side of the road whenever he's more than twenty minutes late getting home. Suddenly, that penchant didn't seem quite so neurotic.

Five days earlier, Dennis had gone home with his friend Samarbek to dismantle Samarbek's family yurt, a sort of felt-covered Central Asian tepee, and to bring their animals down from summer pastures hard on the Uzbek border. It was the kind of escapade we'd dreamt of when we'd moved to Bishkek: riding a horse through a mountain valley hemmed in by jagged snow-capped peaks; listening to old men singing the legends of Manas, the mythic Kyrgyz hero; late-afternoon baths in warm springs bubbling out of the mountains; and learning to assemble, or at least disassemble, a yurt.

We hadn't excluded from that vision sleeping with ten other people in a yurt that stank of sheepskins cured with rancid wheat paste; drinking *kumuz*, the fermented mare's milk that is the national drink; or being forced to toss back glass after glass of rotgut, homemade vodka. Being captured by the IMU, however, hadn't been on that list.

Inflamed by a call from Osama bin Laden to unite all of Central Asia in a single fundamentalist state, the IMU had been kidnapping foreigners and killing locals since the summer of 1999, when rebels had slipped across the border from Tajikistan and raised the green flag of Islam over the tiny village of Zarlady. That first year, they'd traded four Japanese mining engineers and five Kyrgyz officials for several million dollars in ransom before escaping back to safe havens in Afghanistan. The next summer, they'd captured four American rock climbers who'd set up a base camp in southern Kyrgyzstan, the only danger on their minds a snapped bolt in a portaledge or a sloppily driven cam. But in the summer of 2001, a more sophisticated IMU had emerged. Hardened by participation in the Taliban's annual offensive against the Afghan Northern Alliance, they had wreaked havoc across the South, attacking remote military outposts, police stations, television relay towers, casting the specter of a full-scale civil war over the nation.

Just before we'd left home in August, the U.S. State Department had posted a pointed travel advisory designating the IMU a foreign terrorist organization and warning Americans to stay away from the southwestern part of Kyrgyzstan. I'd heard about that warning just as we were draining the water from our pipes in the Catskill Mountains and adding one more bottle of Imodium to our luggage. My friend Frank, a newspaper reporter, forwarded it to me as a good-bye present. His only comment appeared on the subject line, and was limited to two words: "Bon Voyage."

At the time, I'd laughed at Frank's ineffable wit, but as Ambassador O'Keefe ended the meeting with a prayer for our collective safety, I was trapped between the surety that I'd been living with a naive sense of security and the conviction that my fellow Americans were all plunging into a hysterical sense of dread.

On the night of September 11, someone had smashed a few windows at the library at the American University. Soviet Street, where I lived, was defaced with graffiti proclaiming FUCK AMERICA. And I wasn't the only one who had overheard a minibus driver or a stranger on a street corner voice his support for bin Laden, to the hearty approval of some passersby.

But as I trudged to school each day and ambled through the markets, I couldn't find the face of hatred. I saw worry that a flood of Afghan refugees might flee north, washing extremists across the border. I heard fear that homegrown fundamentalists might be emboldened by the fires lighting Manhattan's night. Mostly, I sensed the same resignation that had engulfed everyone I knew, all across the planet, that we were captives to forces we had not yet begun to dissect.

Yet in the back of my mind, a small voice whispered, None of the people who lived and worked next to the terrorists who turned those planes into weapons of mass destruction sensed any hatred either. In the new world arising from the ashes of the World Trade Center towers, the shape of safety no longer seemed clear. I wasn't sure, then, what to think. Like all other Americans, at home and abroad, I had lost the map by which I read danger. A newcomer in Kyrgyzstan still struggling with the difference between the Russian *e* and the Russian *ye*, still unable to read body language or innuendo, I didn't know whether to be afraid, or even what to fear.

I knew only that I had traveled a very long way to a very odd destination. And as it grew odder by the minute, I couldn't help but ask myself a question both embarrassingly trite and singularly apt: Why was I, a respectable American woman with a house, a dog and no demons to flee, always ending up in all the wrong places?

I'd landed in Kyrgyzstan by process of elimination.

That's a pretty feeble explanation in an age when any self-respecting author romanticizes her voyage to a far-off land with a heartrending tale about the aging immigrant nanny who instilled her with the yearnings of the displaced; or the sweet memory of a stubborn childhood fantasy that solidified into an adult promise.

But I never had a nanny nor do I recall having any childhood fantasies about Central Asia, a part of the world that was definitely not on the curriculum of my white suburban elementary school in the 1950s. As usual, the truth is more tediously prosaic: Well into middle age, my husband, Dennis, and I were stuck in complacent ennui and were in danger of riding that rut until we were too old to throw our stuff into backpacks and take off.

Mind you, it was a pretty nice rut, based out of an apartment on the Upper West Side of Manhattan and a home atop a 3,000-foot peak in the Catskill Mountains. I wrote books from an office with a seventy-mile view, and Dennis designed homes and built furniture. No bosses, no clatterous neighbors, no office politics. Just Dennis, a pathologically friendly black lab named Mensch and me on 226 glorious acres and in 3,500 square feet. In other words, we were entirely too comfortable to fit our own self-concept as adventurers.

We needed to shake ourselves up, and the tonic we prescribed was a great journey out of our own culture and assumptions before our luggage became weighted down by arch supports, hormones, pain medication and the rest of the detritus of late middle age.

Financing their own exploits makes journalists queasy, a variation on the professional nervous tic that keeps lawyers from picking up lunch tabs. So I reached back into my former life as an academic for an overseas entree and paycheck and landed on the website of the Fulbright Program. Created in 1945, at a time when Americans were keenly aware of how little they understood the forces that impelled Japanese suicide bombers and German anti-Semites, the program was the antidote proposed by Senator J. William Fulbright of Arkansas, a bold government initiative that would send hundreds of Americans to study and teach in foreign lands. Over the past half century, more than 40,000 American faculty and professionals have worked at universities from Chile to China. I ran down the list of openings, shopping for the perfect country, hoping to become one of the 800 chosen to serve in the Fulbright class of 2001–2.

We dismissed Western Europe out of hand, Rome or Madrid feeling too tame for a couple who had taken their first vacation together in Albania, honeymooned in Havana and still savored the memory of sneaking my research notes for one magazine story past the Mauritanian secret police. We joked that we were saving the first world for our old age, when we would need reliable hot water, mattresses without lumps, and food fried in oil no more than a week old.

South America seemed a logical match since in my first career I'd been a professor of Latin American history. But after thirty years exploring that continent, I was unlikely to be served up many surprises. A Bulgarian job description suited my credentials to a T, but already a candidate for membership in the European Union, Bulgaria seemed as exotic as Boise. And while we were intrigued by China, remote enough from our reality despite the opening of Starbucks inside the Forbidden City, I feared the competition would be fierce.

Mongolia stopped me in my tracks. I'd been fantasizing about Ulan Baator since 1967, when I'd read that the city saw the fewest American visitors—under twenty a year—of any country on Earth. "Are you crazy?" Dennis shouted when I raced down the steps with the news that I'd found our Eden. "Ulan Baator is the coldest capital in the world. What would we DO all winter?"

One by one, we discarded countries: Nepal was a leading candidate for a day and a half, until I realized that the altitude of Kathmandu might not be heaven for a committed smoker. Bosnia and Herzegovina? A recovering war zone wasn't exactly our definition of exotic. Indonesia? Too hot. Pakistan? Too dangerous. Uzbekistan? They were advertising for a specialist in media ethics, a laughable concept since it's tough to be ethical when the government throws you in jail the first time you print anything approaching the truth. There definitely wasn't enough room in Tashkent for my unstoppable mouth.

By the time we were done, Kyrgyzstan was the only country left, and we probably didn't dismiss it because we didn't know enough to come up with any reason to do so. The sum total of our collective knowledge about Kyrgyzstan was that it was a former Soviet republic located over by the rest of the -stans near the Chinese border.

I called all my friends—writers and intellectuals, artists, college professors and publishing types who are supposed to know at least something about everything, or at least enough to make a good show of it. "Kyrgyzstan?" repeated Patrick, who is incredibly well-read although he's a postmodern artist. "Let me look on the map."

"You mean Kazakhstan?" Lisa Bankoff, my literary agent, asked.

"No, Kyrgyzstan."

When she proved incapable of pronouncing the name of the country, I figured that she wasn't going to be much help.

"Kyrgyzstan?" asked my friend Susi. "Are you having a midlife crisis? Searching for your inner child?"

The Fulbright Web page made the country sound deliciously intriguing: "Grand mountains with luscious pastures . . . actively pursuing ethnic tolerance and democratization . . . the most liberal media in the former Soviet Union." And the travelers who write for guidebooks like *Lonely Planet* or post reports on the Internet gave the country rave reviews: "a wealth of nomadic traditions," "Central Asia's finest mountain architecture," "laid-back hospitality."

The more I read, the more convinced I became that Kyrgyzstan was that most idyllic of all destinations: a Shangri-la hardly anyone else had discovered. Bishkek was a "pleasant, leafy, cosmopolitan" city set against a dramatic backdrop of craggy, pristine mountains. Outside town, deep gorges cut into the peaks, a base for hikes to the high glaciers or long soaks in sulfurous hot springs. In the southern city of Osh, Uzbeks and Kyrgyz, Tajiks and Russians and Tatars haggled in a half dozen languages in a thunderous bazaar overflowing with fruit and vegetables, smuggled cigarettes, traditional hats, handmade knives and old Soviet Army uniforms. There was even a beach at Lake Issyk-Kul, thousands of miles from the nearest ocean, at 5,000 feet above sea level.

There's a storybook quality to Central Asia, part Rudyard Kipling, part Genghis Khan, part the isolation from any great— even any not so great—ocean in a world defined by bodies of water. It's laced with the aura of the Great Silk Road, that first link between the East and the West along which caravans of horses and camels, traders and missionaries braved breathless mountain passes and searing deserts to bring gold, ivory, coral and textiles—not to mention Buddhism, Manichaeism and Nestorian Christianity—to China in exchange for furs, cinnamon bark and bronze weapons. (We had no illusions about emulating the great travelers along the Great Silk Road. It took Chang Chien, its first chronicler, thirteen years to make it from China to Samarkand. We didn't have nearly that much time.)

Long before the word *imperialism* was bandied about by Europeans, Alexander the Great had pressed northward from Mesopotamia, so taken with himself that he began dressing like the Oriental despots he sought to conquer. Genghis Khan's hordes rode across the steppes, setting back the flow of civilization in the region for half a millennium. Then Tamerlane—called Timur in Central Asia—rose to power as the first of the great Khan wannabes and massacred tens of thousands in a single day, building pyramids of their skulls.

It was in Central Asia that the Cold War was rehearsed in a surreal nineteenth-century drama between Great Britain, projecting its
own expansionist lust onto the czarist court, and Imperial Russia.
The Tournament of Shadows, as the Russians called it—the Great
Game, in British parlance—was a century-long chess match of high
intrigue, of espionage, paranoia, skulduggery, puppet states, betrayal
and swaggering braggadocio in which Central Asians, Muslim
rebels and Tibetan bandits were bit players in a drama on their own
stage.

Our eyes were also fixed on a past less distant, on the romance
of a nation just emerging from communism, of a people reclaiming
their nationality and identity, and the challenge of helping to train
the first generation born without an Iron Curtain blocking their
view of the world.

What an opportunity! Our lives would be one part Marco Polo,
one part Arabian Nights, spiced with a dollop of memorabilia from
the old Soviet Union and the excitement of working with Central
Asia's emerging Jeffersons and Franklins.

At least that's what I imagined.

Three days after the Hyatt meeting, I woke up, fiddled with the
espresso pot I'd lugged all the way from home and carried my coffee into my office to catch up on my overnight e-mail, my new
morning ritual in a time zone eleven hours ahead of New York.
The first message in my in-box was from the embassy, marked
URGENT TRAVEL WARNING:

> *The Department of State warns U.S. citizens to defer travel to
> the Kyrgyz Republic . . . Americans in the Kyrgyz Republic
> should review their personal security situations and take those actions
> they deem appropriate to ensure their safety, including consideration
> of departure from the country.*
>
> *U.S. Government personnel are prohibited from traveling to
> areas of the Kyrgyz Republic south and west of Osh and in rural
> areas along the Kyrgyz-Uzbek border.*

I still hadn't heard from Dennis, so I flinched at the last paragraphs. But the flurry of e-mails that followed, all from friends back home in Manhattan who were dealing with danger that was more than theoretical, threw the warning into perspective.

I stayed on-line, and was in the midst of reassuring my friend Susi that we were 350 miles from Afghanistan, so, no, we weren't in danger of being hit by an errant bomb, when my computer flashed a message from the warden who was my contact with the embassy. "I've been trying to call you for hours," she wrote, attaching a terse e-mail from the State Department. I scanned the first paragraph, which repeated the earlier advisory, and almost hit the delete button. Then my eyes hit paragraph 2:

> As a result of these concerns, the U.S. Embassy in Bishkek has requested the authorized (voluntary) departure of non-emergency personnel and family members of U.S. Embassy personnel in the Kyrgyz Republic.

I logged off the Internet and started calling my friends.

"So, what are you going to do?" I asked or they asked. Nobody wanted to speak first, to take a position, to admit that he was scared or she thought the whole concept of evacuation was ridiculous. Everyone was waiting for someone else to tell them what to do. Rather than make a decision myself, I fired off an e-mail to the Fulbright office in Washington, begging for some direction, comforted by the thought that they'd make up my mind.

One of the four Fulbrighters, a lawyer from Boston, didn't wait for a response and made reservations to fly home. The Civic Education Project, which sent professors to all parts of the former Soviet Union, informed its fellows that they would be immediately transferred to other countries, and the faculty members from Indiana University teaching at the American University in Kyrgyzstan were recalled to the States. All of the Peace Corps volunteers were told to pack up and go home.

Within twenty-four hours, the undecided began to sway with those winds.

In the midst of the storm, Dennis called from Osh, bubbling

with exotic tales that had nothing to do with evacuations, or even with yurts, meadows or *kumuz*. While I'd been trying to avoid telling the embassy why my husband was mysteriously absent, he'd spent the week being shown off—"Look, a real live American"— and being led on a jaunty tour of the canyon from which the American climbers had been kidnapped and the trails along which the IMU carried their supplies.

"Wait 'til you come back here with me," he babbled excitedly. "You're gonna love it."

Having little talent for self-control, I bit my tongue rather than blurt out, "I hope you took a lot of pictures because it's doubtful I'll ever get there." I'd decided not to mention evacuation until Dennis was safely ensconced in Bishkek.

But as he giddily described his adventures, going on and on, I finally interrupted: "You need to come back as soon as possible. Try to get on a flight tomorrow."

Although puzzled, Dennis tried to comply. Buying a ticket on one of the three daily flights to Bishkek, however, proved a daunting task.

"Everything is sold out until Monday," the clerk at the airport curtly informed him. Three days seemed a bit much, so Dennis pushed, sure that the woman had a few tickets stashed in her back pocket for "special" customers.

"My wife really needs me at home," he lamented, trying for charm.

"But there aren't any seats," the woman said, softening a tad at the considerate husband.

Dennis pushed his passport across the counter, a 200-*som* bill— the equivalent of $4—clearly visible. A quick learner, he'd already mastered the basics of getting things done in Central Asia.

The woman brightened. "Come back an hour before the afternoon flight and I'll see what I can do."

We'd forgotten that Osh was celebrating the fiftieth anniversary of its state university that weekend, so the city was packed with dignitaries whose clout outweighed Dennis's 200 *som*. Dennis had no choice but to ride home in a rickety Lada, fourteen hours up and down the mountains, through ill-lit tunnels, bumping and grinding over roads being rebuilt by crews from Iran. By the time he arrived back in Bishkek, he could barely walk.

Two minutes later, I told him that we were facing evacuation. Washington had kicked the decision back to us, so we were on our own.

It probably sounds mildly crazy that we even considered staying. But we weren't reading the American press or watching American television, so we didn't know that we were supposed to believe that Muslims all across Central Asia were rising up in fury against the godless American heathens. I learned about that terrifying new reality only gradually, in hysterical phone calls, snippets of e-mails, and articles sent by friends convinced that I had lost all semblance of rational thought and needed real information about the place I was living so that I'd wake up and catch the next flight home.

Their arguments had a persuasive internal logic:

In Kyrgyzstan—as in all the other predominantly Muslim areas where the Soviets had extended their control—Islam had been despised, ridiculed and repressed. For seventy years, mosques and madrassahs had been closed, three of the Five Pillars of Islam had been banned, and Koranic courts had been shut down. Then "freedom" came, and Islam was assaulted again, this time by Western movies and music, by Protestant missionaries, by embassies preaching tolerance, and by political leaders more respectful of their bank accounts than of Muslim law.

And that freedom hadn't even brought prosperity. Two-thirds of the population existed below a poverty line set at an appalling $42 a month. According to the Food and Agriculture Organization of the United Nations, one out of every ten Kyrgyzstanis was undernourished. University tuition cost more than the median national income, hospitals refused to treat patients who couldn't afford to bring their own drugs, and tuberculosis was epidemic, the rate three times higher than the average in Latin America or East Asia.

According to the latest recipe—add misery to Islam and stir—Kyrgyzstan should have been on the brink of a fundamentalist revolution, and I should have been heading for the nearest airport since the second course served up by that revolution was an attack on the Stars and Stripes.

I'm not a danger junkie, one of those reporters who gets a rush from pushing into a riot in Pakistan or trailing troops into the Tora Bora caves, so don't think that I was overdosing on the adrenaline of risk. But nothing I'd seen or heard or felt during my month in Bishkek fit into that paradigm of peril. My students, almost all Muslims, wore the shortest miniskirts and highest heels I'd seen east of London, and they didn't seem anxious to trade them in for *borqas*. Practitioners of Muslim Light—what I thought of as the Islamic version of the Unitarian Church—they didn't care about Israel, if they even knew where it was. And they didn't want Americans, who'd just arrived in the early 1990s, to leave them alone so that they could adopt Islamic Shari'ah as law. They wanted us to stay and hand over the keys to a better life. Like have-nots throughout history, their greatest aspiration was to be haves.

"Am I at risk?" I asked them. Who would know better than young men and women from all parts of the country?

"Who would want to hurt you?" they responded, clearly perplexed.

I'd spent enough time on the West Bank to recognize the cold glare of hatred, and in Miami during its murder epidemic to distinguish the pungent odor of danger, so I knew they weren't wrong. There was no rime to my neighbors, no whiff of threat in Bishkek.

Dennis hadn't been home an hour, then, before we conveyed our decision to Washington. "We'll be in touch as the situation unfolds. But, for the moment, we're holding fast."

No matter our pretenses or personal illusions, journalists are, at heart, storytellers, and I admit that stories have a nasty tendency to get messy on me. They refuse to unfold in the orderly sequence hammered into my skull by a dozen journalism professors and scores of editors unmoved by my plaint that the most compelling tales flow to their own rhythm and that the best their tellers can do is to paddle furiously in their wake.

My travels were a lot like those stories, their momentum overwhelming the cheery labors of travel agents, the rigid schedules of

airlines, the most compulsive of planners. I didn't move to Central
Asia to collect experiences for a book. In one sense, I was trying to
dodge the misery of the blank screen, the sinking feeling I always
get when my computer demands a name for a new file just because
I hit SAVE. And I certainly wasn't digging for news, having over-
dosed during the election of 2000 to the point that, like so many
other Americans, I would have gladly installed my plumber at 1600
Pennsylvania Avenue if he would have promised—read his lips—
never, ever, to utter the word *chad*.

I'd looked to Bishkek, then, as a respite from reality, but the nar-
rative I'd planned, like many good narratives, was waylaid by history. The Great Silk Road transmogrified into a trail of terrorism,
Timur was supplanted by bin Laden, the players in the Great Game
Redux shifted on the board, and Stalinism proved once more that
it could outlast Russia and communism. I'd set out on an odyssey
through the ancient lands of Genghis Khan and Vladimir Lenin and
landed at the intersection between the past and the future, at the
corner where Ronald Reagan's Evil Empire met George Bush's
Axis of Evil.

Rather than plod through the expected, I let myself be swept
along past the graceful architecture of old Silk Road cities like
Khiva and Bukhara into the center of Kabul during its first unreal
days of freedom and to an Iran just emerging from behind its
national chador, moving to the syncopation of calculated daring
and the adagio of improvised caution. I dwelled in the world of our
country's old enemies, but I greeted the Muslim New Year on the
streets of Ashgabat, a capital I'd never heard of, and celebrated the
end of a long winter inside an Iraqi airbase at the house of Abraham, in Ur.

By the time I came full circle, after a trek across Russia, Mongolia, China and Indochina, everything I had read and heard about
the shape of America's new reality had been assaulted by experience. I had witnessed a dozen permutations of a rarely discussed
tug-of-war between tradition and modernization—between the
comfort and safety of the old and the allure of the new—that was
heating up into a not-so-cold war. The explosive divide between
communism and capitalism that had held the world captive for half

a century, I realized, had given way to an even more treacherous competition between the yearning for change and the longing for the known. In the lands of Shiite clerics and Muslim fundamentalists, it was playing out as the defense of Islamic purity against an onslaught of excess and corruption. In villages in Uzbekistan and Kyrgyzstan, it had become the struggle to preserve tribal identities. On a vast steppe long proud of its hammer and sickle, it was unfolding as an agonizing attempt to maintain dignity in the face of defeat.

For westerners, this new reality wore the stern visage of the Taliban, the angry grimace of Muslim fundamentalists and bearded terrorists fighting to erect an Iron Curtain against change. But that was but one face of the edgy confusion brewing wherever the tempo of progress—of the convenience of cellphones, the comfort of cars, the seduction of a thousand alien ideas, even the prospect of regular meals—failed to keep up with its promise, or was outpaced by the backlash against its unadvertised costs, a price exacted in the destruction of tradition and daily blows to identity.

When you begin a journey, you rarely understand precisely what you are seeking—and in any event, if you look too hard for it, you fail to see what you have found. Beneath the flashing neon lights of Tehran, I met mullahs on motorcycles convinced that modernization was a pact with the devil. In the classrooms of Kyrgyzstan, I taught students repulsed by American materialism, yet looking for visas to work in New York. No matter where I landed in Iraq, I was welcomed because I was American; in the former Soviet Union, I was reviled because of the Winter Olympics.

Out of that jumble of envy and confusion, of the thirst for change and the hunger for continuity, came a year of e-mails to friends back home. From those e-mails emerged a book.

—Elinor Burkett
Old Clump Mountain, New York

part ONE

Forward into the Past

No matter how often you say Halvah,
the taste won't appear in your mouth.
—RUSSIAN PROVERB

There's a moment in any voyage, just after I clear customs, when I pause in the netherworld between the timeless, placeless, plastic surreality of the airplane and my chimera about what's to come. On the one side is the familiar, the routine; on the other, the door to the unknown. It's a delicious interlude, fantasy ruling for one final instant. Then the doors swing open . . .

On August 25, 2001, this is what I saw:

A churning sea of Oriental men smiling and screeching, blocking the exit, turning my first glimpse of life in Kyrgyzstan into a lesson in not-so-diplomatic assertiveness. Taxicab drivers, porters and shills jockeying for meager tips pulled at my jacket, grabbed at my luggage, bargaining—with each other, not with me, because I was too busy clenching my teeth to dicker. "*Pitisot somof*, 500 *som*," yelled an older Kyrgyz man in a worn leather jacket. "*Nyet, chitirista somof,* 400 *som,*" a young Russian offered, shouldering the other man aside even as a third man caught my eye, pointed to himself and whispered, "*Uch juz, besh juz,*" which I assumed was a price in Kyrgyz, although 300 *som* was unlikely to sway an obvious foreigner unlikely to understand the language.

Nyet was one of the few words I knew in Russian, and it was the only word that would come out of my mouth.

Nyet to the drivers.

Nyet to the porters.

Nyet to the shills.

I was waiting for the embassy driver and Nurilya Barakanova, the public affairs assistant, who should have been the "woman with the cellphone." But everybody had a cellphone, and since the drivers wouldn't take *nyet* for an answer, the price of a taxi had fallen to *dvesti somof,* 200 *som.*

It was 4 A.M. in Bishkek, 5 P.M. back in New York, which I had left forty hours earlier. Outside it was pitch-black, although it wasn't much brighter inside the terminal, where half the bare light sockets in the ceiling were empty, the other half filled with yellowed forty-watt bulbs. The only store open was a small kiosk run by a woman so surly that passengers would have had to be desperate for gum or a cup of tea for her to make a sale.

We finally found Nurilya, tall, slim and stylish, standing with her cellphone to her ear, loaded our baggage into the white Toyota Land Cruiser and headed off for the city. Somewhere along the lovely tree-lined highway from Manas International Airport to downtown, I blanked out, my mind on overload, my body demanding relief after a four-hour drive to the airport, a seven-hour flight from New York to London, a six-hour layover and the ten-and-a-half hour trip to Bishkek—not to mention a sudden mind-numbing what-have-you-gotten-yourself-into panic. When I awoke twenty minutes later, we were on an obstacle course, the challenge being to circumvent at least most of the cavernous potholes that made the road feel like macadamized Swiss cheese. Our driver was not meeting it.

As I looked around me, I saw the outskirts of the city, row after row of tightly packed, tiny Ukrainian-style wooden houses washed

with thin, streaky coats of green, brown or blue paint. In the neighborhoods behind them, dirt tracks were lined with high fences
constructed of wood, wire, steel and trash. All the shutters were
tightly secured, no light escaping to suggest the presence of a night
owl cozily prodding himself to sleep with warm milk.

The city itself appeared abruptly, the street suddenly transformed
into the floor of a canyon created by block after block of stolid
apartment buildings, all of Soviet gray cement, the near-universal
shape and color of Eastern Europe and Central Asia, from Warsaw
to Almaty. The only relief from the sea of cookie-cutter structures—built from prefabricated concrete forms in one of the eight
approved patterns—were the balconies, seemingly the sole design
element over which an architect had exercised some creative control. Some were formal and rectangular, traditional, practical. Others were rounded or elongated, the most recent even embellished
with what looked, in that context, like daringly bold, almost avant-
garde pop art concrete designs. But each balcony had become pris-
oner to its owners' proclivities or financial means. They were
bricked in, boarded up, sealed with corrugated fiberglass, windowed
or set off with fancy railings. Laundry hung, detritus collected and
sick-looking plants languished.

Even as dawn broke, Bishkek was dark, lit only by the infrequent
streetlight that had not blinked itself out of existence in a city
where burned-out lightbulbs were only rarely replaced. Just down
from the Grecian-columned Opera House, the Hyatt Regency
Hotel and the round Wedding Palace that were the heart of "official" Bishkek, the driver wended his way through the gloom into
an alley between two crumbling brick buildings, ending up in a
courtyard bound on one side by an old red-brick apartment building, on the other by a row of padlocked metal garages so rusted
that they no longer looked like they had ever been gray.

As Nurilya urged us out of the plush Land Cruiser, we trudged
past a jumble of reeking garbage cans straight into the smiling
countenance of Alexander Katsev, the head of the Department of
International Journalism at the Kyrgyz-Russo-Slavic University,
who escorted us through a cold metal security door that screamed
the South Bronx.

I'd been dreaming of my first meeting with my new boss for five months, but I hadn't expected it to occur at 5:30 A.M. or to be scheduled so that he could try to convince me to rent his apartment.

When we'd made our way up a flight of unlit stairs, through another security door and into a breathless, airless apartment, his wife, dressed in what looked like one of June Cleaver's old shirtwaists, gracefully swung open her arms to usher us from room to room as if she were Deborah Kerr in *The King and I*, or an Upper East Side realtor showing off a penthouse overlooking Central Park.

The rooms were taller than they were wide, suggesting coffin more than home, and most of the width was taken up by the kind of dark, heavy wooden furniture that my Russian immigrant grandmother believed to be classy, or *klessy*, as she pronounced it. The backs and arms of the stuffed couches and chairs were festooned with antimacassars, and the closet-size bathroom was a maze of pipes. The apartment was spotless—immaculate, in fact. But the entire place was painted the color of tobacco-stained teeth.

We smiled politely, admiring the bookcases, the books, the draperies, the overstuffed couch, anything we could think of to keep our mouths from spitting out, "This is the most dreadful apartment we have ever seen." As we entered the kitchen, Dennis relieved his tension by whispering through his teeth, struggling to maintain a grin fixed on his face, "I'll die if I have to spend even one night here."

Nurilya managed to extricate us with some blather about embassy security needing to approve the building, but Mrs. Katsev wasn't fooled, or was very deft at masking her disappointment at our lack of enthusiasm. As we raced out into the dawn to the "temporary apartment" Nurilya had arranged, I wondered if I'd poisoned my relationship with my chairman before it had even begun.

Months earlier, I'd tried to establish some rapport with Katsev with a brief introductory e-mail. What courses would you like me to teach, I'd written, and how, beyond lecturing and seminars, might I assist your program? Are there materials you would like me to bring from the States?

His response should have been an omen, or would have been if I'd known how to read it:

Mrs. Elinor!

We'd like to know, what courses and subjects do you want to conduct? How many hours do you want to work per a week? What job besides lectures and seminars do you like to conduct with students? We hope, all the materials you'll bring, will be very useful for students.

Before I could dwell further on the meaning layered into that e-mail, or how Katsev might repay my refusal to accept his hospitality, at $400 a month, we needed a place to live. Noting our horror at the apartment we'd just seen, Nurilya promised that our temporary accommodations would be a step up from Katsev's abode.

Dominating the corner of Moscow and Soviet Streets, the commercial heart of Bishkek, Dom Torgovli was what passed for a prestigious address. A giant U, the complex of three nine-story buildings had been constructed back in the heyday of communism for foreign guests, party officials, senior military officers and others whose status gave them a right to space in a perfect location with the convenience of an arcade of shops on the ground floor and a playground hidden inside a tree-lined courtyard.

Two sides of our temporary home were lined with windows that gave way on to long, glassed-in balconies. The 1,500 square feet of space was divided into two bedrooms, a huge living room, two bathrooms, one complete with bidet, and a kitchen that included an espresso machine. It was safe and secure, access to the building controlled by a coded security lock, and entrance to the apartment itself by a gated door. The furniture was new, the telephone line was digital and the television/VCR was hooked up to cable.

Or at least that's how a real estate agent would have described it if there had been any real estate agents in Kyrgyzstan. He or she wouldn't have been lying. What she would have been omitting was: The concrete steps into the building were crumbling so badly that the first step had been shored up with a broken fragment of cinder block. The blue paint in the unlit hallway was peeling and heavily graffitied. The elevator hadn't been maintained in twenty years, the narrow balconies blocked much of the light coming into the apartment, the smell of sewer gas wafted out of one bathroom, the ceilings were cracked, and the wallpaper was peeling. The refrigerator had a freezer the size of a shoe box, and the oven had no thermostat. Across from the entrance to the building, the garbage room spilled its contents out onto the sidewalk. Most of the arcade was empty. And the playground, with swings crafted out of rebar, would have made the outdoor space in a Chicago housing project look homey.

For Bishkek, that wasn't a bad list of defects.

When Kyrgyzstan opted to follow the road toward capitalism, housing was privatized and all residents given ownership of their current dwellings. Those who lived in nice apartments quickly discovered they could make a fortune by moving in with a cousin or brother and renting their homes out to foreigners. Those who'd been living in slums when privatization occurred, of course, had no such option, which pretty much etched into stone the class system that existed under communism—except, of course, there wasn't a class system under communism. As we'd already learned in our four hours in the country, it was called something else, like rewards for the peoples' workers or those who dedicated their lives to the Party.

The owner of the Dom Torgovli apartment, a tall, slim Kyrgyz woman with the haughty air of the privileged, greeted us at the door and immediately assumed her position at the head of the dining room table. Having won that capacious apartment, Vilena must have been dedicated to something about Soviet life, but it was clearly not the improvement of the masses. She owned a car, imported furniture and Italian clothes, and had already rented out one of her other apartments to another foreign couple for $500 a month, a princely sum in a city where apartments could be found for as little as $50. Nurilya had agreed that we'd pay $50 a night for our temporary haven, but Vilena offered to move out for the full year if we would give her $450 a month, entirely in advance, which meant that her monthly income in rent alone would be three times the average annual income in her country.

We were hot and tired and had just lugged a year's worth of clothes, books, papers, cameras, medicine and computer supplies up four flights of steps since the elevator, which could not have accommodated four people even if they were extremely intimate, was being balky. We were willing to live with a leaky toilet, the ever-present sewer gas, and lace curtains hung incongruously two feet away from the wall, even at $450 a month, if everyone would just leave us alone.

Vilena was not easily removed, even once she had two thousand crisp U.S. dollars in her hands. First she inspected the twenties, bill by bill, to make sure they were pristine—even the smallest tear, the tiniest crease, made a bill unacceptable. Then she counted them, twice, to make sure she wasn't missing one. Then she wanted to talk about maid service, which she offered us for $50 a week, which we suspected was approximately ten times what the maid would receive. Bread had to be broken, an unbending Kyrgyz tradition, the silverware had to be tallied, and Vilena had to command her two sons and a maid as they bustled around obediently, packing clothes, sorting through papers and picking through the kitchen.

By the time we locked the door behind them, it was after 11 A.M. Exhausted, I walked out onto the balcony to examine the view. Dennis stayed behind in the bedroom, his head in his hands, rocking back and forth.

As promised, the southern landscape was dominated by a high range of jagged snow-capped mountains and, as the guidebooks had advertised, the streets were lined with poplars and birches, opening to a plethora of pristine parks. But Bishkek's planners had turned the city's back on the mountains, which were barely visible even from major intersections. The balconies of the apartments, including ours, faced away from them, opening to views of empty lots filled with trash, noisy streets and the backyards of nearby buildings. Although the greenery was stunning, the leaves, we knew, would fall within six weeks, revealing the shabbiest, ugliest city we'd ever seen.

In places like Peru and Albania, I'd seen some pretty grungy neighborhoods where raw sewage ran through the streets and houses were constructed out of packing crates. But, invariably, across town, the privileged lived behind high gates in something resembling the first world. A commercial area always tried for a veneer of modern. And even on the meanest of streets, wrought-iron railings on old houses curved with whimsy or a handmade pot graced a front door.

Bishkek, however, was more like the deteriorating rather than the developing world, a post–first world, if you will, the first world after the apocalypse hit. You could sense that things had once functioned—that the streets had once been decently paved, the buildings painted, the trash collected. But that past was as distant as the days when Harlem was an upscale neighborhood. Even before neglect became the dominant theme of the urban decor, no time or money had been wasted on frivolities like quirky railings. No bright colors broke the dreary panorama of grays. In the workers' state, the only decoration permitted were the statues of heroes from what people thought of as a glorious past, dominating a not-so-glorious present: Lenin and General Mikhail Frunze, a local boy who achieved glory as a hero of the Red Army; of Felix Dzerzhinsky, organizer of the Soviet secret police; of Manas, the mythic Kyrgyz hero, riding his magic steed, Ak Kula.

When you find yourself freaking out in a foreign country, gasping for air, convinced you'll never regain your equilibrium, go look for the familiar. That's been my mantra for decades, so I shook Dennis out of his stupor, pulled out a map, and we headed for Europa Supermarket, purveyors, I'd been told, of Hellmann's mayonnaise, my symbol of the familiar.

We locked our front door. We locked our security door. We locked the gate that protected us from the hall, and made our way onto the street gingerly—very gingerly, since the steps of Bishkek inexplicably varied from four inches to a foot in height and eight to eighteen inches in depth, the bane of foreign residents foolish enough to rely on standardization. They were an accident lawyer's still-undiscovered dreamscape.

It was Sunday afternoon, and the streets felt lazy, couples strolling to the market, an occasional drunk passed out on the sidewalk. No one else seemed to pay any attention to the rusting cranes sitting idle next to half-built high-rises that had been abandoned ten years earlier. In the parks, city workers looking like grim reapers cut the grass with huge scythes, and uniformed women swept the walkways with brooms that were simple bundles of twigs. Without so much as a downward glance, old ladies, their heads wrapped in the kerchiefs that gave them the name babushkas, weaved around open manholes, their covers stolen, we'd been told, by junk-metal dealers.

As we wandered through downtown, one thing became abundantly clear: we wouldn't be spending much time shopping. Tsum, the old Soviet department store, was still the liveliest enterprise around, two floors of Chinese kitsch, French nail polish, faux crystal, Korean televisions, Russian dishwashers, Japanese computers and eighteen stalls filled with souvenirs for the hordes of tourists who had not arrived. The aisles were jammed with old women admiring glassware, young men playing with VCRs, girls giggling by the makeup counters. But few bargains were to be had at Tsum, and in a country with an average annual income of $290, bargains were everything. The crowds were window-shopping, then, the hobby of people still unaccustomed to the concept of variety.

A furniture store up the street offered ten or twelve pieces of the type of modern Italian furniture that delights the eye but not the back. At the clothing store next door, you could choose between pink, green or yellow blouses, and we passed one shop that sold shoes and another that carried cosmetics. Other than Tsum, tiny markets and vodka shops, that was about it for stores in Bishkek.

On the street, however, the possibilities were boundless.

Just below the plaza fronting our apartment complex, a dozen cake ladies had staked out their turf on a patch of dirt between the sidewalk and the street. Each woman guarded her own pile of flimsy cardboard boxes, most carefully tied shut, a few lidless to catch the interest of passersby and the soot from the belching buses. Their creations were elaborate fantasies in shortening and sugar: hearts outlined with nuts, diamonds with piped cross-hatching, swirls and leaves and fleurs-de-lis dispensed for $1 to $2. When no

customers were near, the women squatted on piles of bricks, gos-
siping and laughing. But as soon as a potential buyer so much as
glanced in their direction, the amiable coffee klatch atmosphere
dissolved into a frenzy of pushing, pointing and pleading, each ven-
dor thrusting a sample of her pastry skill into the intrepid buyer's
face.

Like the cake ladies, the flower ladies, an older, more wizened
group, also congregated. But when shoppers came along and examined their tightly bound bouquets, virtually strangling in cello-
phane, they maintained an almost eerie, noncompetitive
indifference.

At almost every intersection, the water ladies, wearing frilly
white aprons and light blue dresses, sat on low stools next to old
soda vending machines, two or three glasses lined up on tables next
to them. For three cents, they'd fill up a glass with water from the
machine, which they refilled daily with buckets of tap water. Their
principal competition were the Shoro ladies, who dominated the
corners the water ladies hadn't already claimed, selling, for six cents
a cup, a fermented drink that tasted like wheat flour dissolved in
water with a dash of pickle juice.

The weight ladies perched on chairs next to ordinary bathroom
scales: one or two *som*, two or four cents, to read the bad news. The
popcorn ladies were, perforce, a higher breed, since they needed
actual equipment. We fell on the first one we saw with delight,
since for Americans, popcorn falls into the comfort food group.
Alas, instead of filling the salt dispenser with salt, they loaded it
with sugar.

There were ice cream ladies dispensing half-filled cones, and
gum and cigarette ladies peddling their wares by the stick and the
fag. The used-clothing ladies—widows bitter that the socialist
promise of lifelong security had turned into a pension of just nine
or ten dollars a month—displayed old mink collars, ancient Soviet
Army caps and other remembrances of better days on plastic sheets
on the sidewalk. The knitting ladies sold their caps and socks with a
show of skill, purling away as they waited for clients. The lottery
ladies, in bright yellow plastic vests, manned umbrellaed tables as
they sold the chance to win BIG MONEY. The pill ladies hawked

expired medication and small packets of herbs. And the fortune-teller ladies read palms and cards, offering up good news or bad news according to how much you paid them.

Despite the grim backdrop, the honking Ladas and the Soviet uniformity, I caught a few glimpses of the Silk Road. There was nothing ancient in the city, which had grown out of a nineteenth-century Russian outpost. The oldest part of town was its name, Bishkek, a play on the Kyrgyz word for the wooden churn used to make fermented mare's milk. The moniker had been reclaimed after independence in 1991; the old name, Frunze, was imposed by the Soviets in 1926.

The open-air markets, however, were infused with the odor of *plov*, a rice dish spiced with dollops of sheep fat. We walked down aisles filled with newly plucked chickens and piles of singed goat heads, the teeth still grimacing behind curled burned lips. No matter where we turned, we confronted a riot of color: counters lined with bowls of mysterious spices in hues of blue, orange and red, and a hundred varieties of apples, raspberries, pears, melons, grapes, apricots and figs.

And the faces took me back, a melting pot of Mongol and Chinese, Turkic, Middle Eastern and European. The ethnic Russians were the easiest to spot, solid and fair-skinned with an abundance of poorly dyed, gaudy red hair and forlorn countenances. Before the breakup of the USSR, these descendants of the obstreperous who'd been forced south by Stalin and of the land-hungry lured to Central Asia by Khrushchev had made up more than 20 percent of the population. But when Kyrgyzstan became independent and chose Kyrgyz as its official language, they'd fled back to Russia by the thousands. In the first three months of 2000 alone, 13,000 Russians had departed Kyrgyzstan, and there was no sign the flow was stanching.

Older Kyrgyz men stood out at a glance, their tall white felt hats, *kolpaks*, a dead giveaway. But even with a guide, we suspected we'd never learn to tell the difference—so obvious to Central Asians—between an Uzbek and a Kyrgyz, a Tajik and a Kazakh, a Uighur and a Dungan.

All that was missing was a single recognizable Muslim—in a

country with 4 million Muslims, 75 percent of the population. Older women trudged down the street, bent and lined, wearing what looked suspiciously like bathrobes, while young women flaunted their bodies in the same skin-tight jeans and short skirts that are the global uniform of their generation. Except for the few wearing *kolpaks*, all the men were hatless. We saw not a single head scarf, veil or embroidered Muslim cap in the city.

On Ala-Too Square—still called by its old name, Lenin Square, an insurmountable fixation since the old man in the overcoat, whose image had been relegated to junk heaps in most of the rest of the former Soviet Union, still presided over it—workers were draping buildings with flags and setting up reviewing stands in preparation for Independence Day, the tenth anniversary of Kyrgyzstan's reluctant separation from the USSR. Kyrgyzstan had not so much declared independence as been forced into it when the presidents of Russia, Belarus and the Ukraine dissolved the Soviet Union without bothering to consult the Central Asian presidents. Five months before, in a popular referendum, almost 90 percent of Kyrgyzstan's electorate had voted to preserve the union, convinced they could not go it alone.

I wasn't sure how many people would be celebrating anything more than a day off. Since the collapse of the Soviet Union, the Kyrgyz economy had contracted by almost half, and the country was moving backward. Roads had been closed rather than maintained, heat and hot water abruptly disappeared, and families that hadn't been nomadic for generations were taking off to high pastures to live in felt tents.

Our senses assaulted, we took refuge in Europa, where a dozen uniformed shop girls waited for a customer—any customer, even one just buying milk. At $8 for a wedge of Parmesan cheese, $5 for a small container of peanut butter and $3.50 for a tiny jar of Hellmann's, they didn't have much business.

We trudged home in the waning light of the day through a city with all the charm of Stalin's mausoleum, insensate. Dennis muttered in a flat mantric tone, "I don't know if I can do this. I don't know if I can do this." Wondering how Americans had spent half a century terrified of a country where people living three blocks from the parliament still had no indoor plumbing, I clutched my mayo like a talisman.

We tried curling up on our couch that night, looking for distraction in any of the sixty-two channels on our cable. But neither American movies dubbed in Russian, pop festivals from Moscow, reruns of President Askar Akaev's speeches in Kyrgyz, romances from India, news from Korea, or endless commercials from China could divert us. We took to our bed, huddling together on its incredibly lumpy mattress, praying that we were just tired, cranky and stupefied by culture shock, that in the morning, things would look better.

2

Generation Я

We may take long to harness a horse,
but we ride really fast.
—RUSSIAN PROVERB

I didn't know how to react—whether to sit or bow or remain standing—when eighteen students scrambled to their feet the moment I entered a classroom the size of a Ford Expedition on the first day of the fall term at Kyrgyz-Russo-Slavic University, KRSU.

I'd been warned that my students would be deferential to authority. But I didn't catch even a flicker of the sour grimacing or impatient foot shuffling with which American sophomores would undercut any such required gesture of subservience. The slight young man with the blond ponytail didn't so much as slouch. Neither did Dinara—who I eventually learned was considered too "American"—shrink from paying homage to the new professor.

I hadn't moved to Central Asia to bask in some exalted professorial ego trip, but to help train the first generation of post-Soviet journalists to be as dogged and daring as Christiane Amanpour or

Bob Woodward. So I dismissed the exaggerated courtesy as ritual sham, a public display of obeisance disguising predictable adolescent contempt, fumbled with my papers and moved on to the lesson of the day.

My first task was to test the students' journalistic skills, and I'd chosen as my arena the hot gossip in Kyrgyzstan—that the daughter of President Akaev had separated from her husband, who happened to be the son of the president of Kazakhstan, on Kyrgyzstan's northern border. Imagine Chelsea Clinton married to Tony Blair's son and you'll have some sense of the dimensions of the potential scandal.

"How would you find out if this is true?" I asked.

Not a single face lit up the gloom. It wasn't always gloomy, but that day, the first of many, the electricity wasn't working on campus.

"What about calling the president's spokesman?" I suggested, opting for the prosaic. The students' expressions revealed none of the incredulity they later admitted they'd felt at the silliest question they'd ever heard.

The young man with the ponytail, Artyom, took pity on me. "There's no reason to call, because he wouldn't tell us," he said matter-of-factly.

"How do you know if you don't try?" This time, they gawked openly in puzzled disbelief that I read as a clear, Are all Americans as stupid as you?

Okay, I thought, looking around me at the rickety chairs and a blackboard so chipped that it no longer held chalk, so they were unwilling to press those in authority.

I forged on: "How else might you discover whether this rumor is true? Is there a less direct way to find out?"

Sneakiness apparently wasn't part of the journalism curriculum. At least it hadn't been.

"Do you know anyone who works for the Akaev family? Who does Mrs. Akaev's hair? Is your next-door neighbor related to her chauffeur? What about nosing around to find out?"

Dinara, the only girl wearing a baseball cap and Doc Martens instead of high heels and a skirt, laughed.

"This is Kyrgyzstan," she instructed. "You don't ask questions like that in Kyrgyzstan!"

Teaching overseas, especially on the taxpayers' dollar, is a tricky business, since the line between pedagogy and fomenting revolution—an activity the American government would surely discourage, especially in a Muslim country—isn't exactly etched in stone. Being culturally sensitive is fundamental to Intercultural Communications 101. But how do you respect local traditions when you're a journalism professor and those traditions include censorship, self-censorship and pretty straightforward journalistic dishonesty?

It was my first moment of truth in the classroom. No matter the continent, if you blow it the first day by pushing too hard, or not hard enough, you're dead in the water. Young children have mercy; adolescents have none.

"Why not?" I inquired.

"I don't know," stammered Dinara. "It's just our tradition."

I knew I was supposed to stop, but I couldn't. I rarely can.

"Well, does that tradition serve you well? Does it produce good newspapers and a good government?"

Artyom interrupted, "Asking questions like that is an invasion of the president's privacy." Twelve voices joined in, all supporting the president's right to privacy.

The night before, I'd watched *An American President* on television—watched, nothing more, since Michael Douglas had delivered his impassioned defense of Annette Bening in Russian. I was tempted to change gears and to see what my students, so protective of their president's privacy, made of the movie. Too soon to distract them, I decided in a flash. Once students figure out how to lead you astray, you've lost all semblance of control.

"How much privacy and under what circumstances?"

The students quieted.

"Is that right absolute? What if he's having an affair with the wife of the president of Kazakhstan? Would writing about that be invading his privacy or telling your readers something they might need to know?"

They hesitated, and then reverted to, "It's just not our tradition."

I reverted as well, to my original inquiry: "Well, does that tradition serve you well? Does it produce good newspapers and a good government?"

The students didn't know how to answer because no one ever asked for their opinions, on that or any other subject. In Soviet times, young people had studied the ideologically acceptable and the socially useful. They deferred to their professors and, at exam time, regurgitated their lecture notes. Disobedience—which encompassed everything from questioning and probing to expressing an original thought—was a sort of Communist venial sin.

In theory, that system had succumbed a decade earlier with the collapse of the USSR and the creation of a democratic state. But like all habits, old traditions die hard, even in the face of Britney Spears and MTV, the Internet and a democratic constitution. Young people, then, were still conditioned to believe they didn't know enough to have opinions, that they didn't have the right to question, that they had no choice but to bow passively to authority.

My students had opinions, of course, but I sensed I'd need an excavator to unearth them.

"Are you happy with the quality of your government?" I pressed on. "Do you like President Akaev?"

Dinara and Artyom were the first in their group to speak, as they always were. Dinara had learned spunk during the year she'd studied in the United States. Artyom, an intense young Russian who'd made a conscious effort to overcome crippling shyness, was self-taught. Or perhaps he'd inherited a disobedience gene from his parents, who had defied all Soviet attempts to wipe out religion by joining the Russian Orthodox Church.

"No one is happy with the government," they said, almost dismissively. "But it's the government. It's not our right to ask questions of the government."

Let's see what the rest of the population thinks, I said, explaining that I planned to teach them how to conduct at least a semi-scientific poll.

"What's the purpose of a poll?"

"In the West, we use polls to gauge public opinion."

They remained perplexed. "But why?"

"To tell the government what the people want."

"But why do they need to know that?"

I had an hour between classes, so I headed outside and plunked myself down on a stoop at the bottom of the unused grand staircase that fronted the building to figure out how the hell I was going to teach journalism to such students. The scene on the terraced plaza before me didn't look much different from campuses in the United States, Europe or any other continent. It was a perfect beginning to the fall term, the air crisp, a hint that snow would soon creep down the mountains toward the city. Students were smoking cigarettes, checking out the freshmen, yelling in delight at reunions with friends who'd been away for the summer, already complaining about their professors. Kyrgyz chatted with Uzbeks who were dating Koreans and hanging out with Chinese. But the Russians outnumbered them all. Established in 1992 as a sop to Russian Kyrgyzstanis, KRSU was the last bastion of the old Russian intelligentsia.

The only jarring element was the sense of style. No baggy sweatshirts or unpressed jeans, no piercings, pink hair or shorts. The young men wore jackets and the occasional tie, the young women preened in perfect makeup and lots of hairspray. Unabashedly Penn State, class of 1955.

I was too wound up, too disconcerted, by my first class to take in much of the scene. I was in over my head and, what was worse, I knew it. Using Mel Mencher's textbook, the sine qua non of journalism education in the United States, I thought I might succeed in training my students in the basic skills of interviewing and writing. But it was clear that they would be loath to practice them.

My first official meeting with Katsev should have prepared me. "What theory do you use to teach journalism?" Katsev had asked me. Like most local journalism professors, Katsev had never been a journalist. He was a philologist, a scholar of the Russian classics, and his notion of training journalists was to encourage students to pen essays or rewrite government press releases with the elegant prose style of Alexander Pushkin, Russia's most famous writer.

I was seated at what he had offered as "my desk," one of four in the departmental office, which also contained two secretaries, one telephone and a steaming electric kettle. No office in Bishkek could exist without a means to make tea. Katsev alternately paced—not an easy task in a ten-by-twelve-foot room with little open floor space—and leaned against a secretary's desk. He didn't walk through the dark hallways of the building; despite a pronounced limp, he raced, his ruddy face ruddier still, his unruly reddish hair wafting in his wake.

I'd never had a class in journalism theory, and I admit that if one had been offered at the Graduate School of Journalism at Columbia University, where I studied, I would have viewed it with the distaste I usually reserve for corned beef hash.

"I don't believe in theory," I responded candidly. "I believe in practice."

Katsev was inclined toward the baroque in his language, which was laced with old Russian aphorisms, quotes from Russian poets, both famous and unknown, and long, complicated tales from the lives of Russian literary figures that always turned out to be morality plays. Raised in a Russian Jewish household in Philadelphia, I knew the type well. They talked you into submission: death by verbal drowning. He was gushing faster than Niagara.

Fortunately, I could understand none of Katsev's response because he couldn't spew forth in English, and I understood just enough Russian to own up to the fact that I spoke no Russian. Three translators were on hand—two students and Nurilya from the embassy—but they didn't even attempt anything approaching a literal translation of the harangue that followed, boiling it down to a single sentence:

"There is no practice without theory."

A child of the 1960s, I'd studied my fair share of Marxism in college.

"Lenin's been dead since 1924."

The gist of Katsev's reply:

"Certain truths are eternal."

The gist of my response, offered gently but unmistakably:

"Not in my classroom."

It had been yet another inauspicious beginning to our relationship.

Katsev's primary concern was that his students not be encouraged to practice "yellow journalism," which he was convinced had established a foothold in Kyrgyzstan. I wasn't sure what he meant by the phrase, an American concept that refers to the salacious and flamboyant scandalmongering that transformed urban newspapers during the late nineteenth-century. Bishkek's newspapers, filled with long political essays, literary allusions and abstruse economic analyses, were a far cry from the punchy publications of Joseph Pulitzer and William Randolph Hearst that had given rise to that epithet.

Most days, the closest *Vecherny Bishkek* ("Evening Bishkek") or *Res Publica* ("Public Matters") came to salacious were five sentence blurbs announcing that the rate of HIV infection had doubled or that prostitution was on the wane. When they produced real scandal—by accusing the president of maintaining a villa in Switzerland or a judge of taking bribes, usually with minimal evidence—the reporters and editors tended to wind up arrested for criminal libel. The owners of opposition newspapers were regularly harassed by the tax police, sued into bankruptcy and visited by the MNB, the new name for the old KGB. You can imagine, then, how common hard-hitting reporting had become.

I suspected that Katsev's real fear was that I might teach my students to rock the boat. Boat rocking was not on my syllabus. But reporting, a skill my students seemed not to have learned from any of their regular professors, certainly was—and I feared Katsev would see no distinction between serious reporting and untoward provocation.

In the United States, aspiring journalists study the basics, from math and chemistry to political science, then learn their craft in courses like Reporting and Writing 101, Advanced Reporting, Investigative Reporting, Computer-Assisted Reporting, Science Writing, Magazine Writing and Political Reporting. They read daily newspapers, break down articles to examine their structures, sift through municipal budgets and pick apart press releases. Then they go out on the streets—on campus, to city hall, to local hospitals and police stations—to produce stories of their own.

In Kyrgyzstan, aspiring journalists also studied the basics, with a heavy emphasis on literature, and then they learned their craft in courses like Psychology of Journalism, Theory of Journalism, Theory of Creativity in Journalism, History of Journalism, and History of Soviet Journalism. They parsed out Shakespeare, Thornton Wilder, Leo Tolstoy, Fyodor Dostoyevsky and Chinghiz Aitmatov, Kyrgyzstan's most famous writer. They were not required to read public documents, the few that were published, and they never, ever ventured off campus to practice reporting. The closest they came to committing journalism was during brief practica between semesters, when they were sent to "real" journalists for experience.

I hadn't moved from New York to Bishkek to teach Theory of Creativity in Journalism.

"Excuse me, but you must not sit there." A young blond Russian in perfectly pressed black pants and a clingy, fluffy pink sweater interrupted my train of thought. I'd been so lost in academic despondency that I hadn't seen her approach. Looking up, I realized that half the students gathered on the plaza were staring not-so-furtively in my direction.

"Am I doing something wrong? Is it prohibited to sit here?" I asked. I suddenly realized that no one else was sitting.

"Maybe as foreigner you don't understand, but sitting on cold cement is bad for making babies," she said earnestly, as if instructing a primitive. Her tone was precisely what I would use if I saw, say, a Bolivian peasant swimming in the East River.

Dennis had discovered the local phobia about cold genitalia two days earlier while sitting on a wall in the park. A young man had motioned him to get up, shaking his head in admonishment, clarifying his curious behavior by pointed to his groin and curling his index finger, as if miming a flaccid penis. We hadn't realized that the concern was not gender-specific.

When you live abroad, some voice of conscience or guilt, or the old saw "When in Rome . . . ," inclines you toward local custom. But local custom was to hunker, a balancing act my untrained Western legs could not master, or to stand. Still dizzy from jet lag, I stayed put.

"Thank you, but I'm a bit too old to have children in any case," I explained. My instructor in Kyrgyz customs nodded her head, not a whit of comprehension relaxing her face, and rejoined her friends, who continued staring at the new professor, the American, clearly a novelty at KRSU.

I still had five minutes before I needed to begin the long trek to my classroom, a labyrinthine hike through the main building, one of just three that were the "campus" for 4,200 students. Its imposing colonnaded facade bespoke its former glory as a headquarters of the Ministry of Defense, the only sign of its reincarnation a statue of Pushkin, a firm declaration of the university's ethnic loyalties. The architectural style was Soviet intimidation, the interior a maddening maze that no Chechen rebel would have dared hazard. The main entrance under a flared arch at the top of the wide staircase was chained shut. Hidden beneath that stairway were three sets of exterior doors that opened onto three more sets of exterior doors. Some days the first right-hand doorway was open, while only the inside left-hand doorway permitted access to the building. Other days, the entry code was reversed or revised or configured in yet another combination. Prisoners of paranoia, either about an assault or a gust of wind refreshing the stale air, the keepers of the keys refused to unlock all the doors, creating a permanent bottleneck whenever a class bell rang.

Those keepers were Russian security ladies, hefty and humorless, who controlled access to all the rooms in the building from a battered table inside the lobby. They didn't need guns or bats to keep out the unwanted. Their disapproving grimaces were menacing enough. One glance warned you that forgetting to return the keys you needed to sign out in order to open your classroom would be a grievous error.

The lobby was dominated by a grand staircase to the second floor, the kind Scarlett O'Hara would have confidently sashayed down but for the overstuffed brown Naugahyde that padded its railings. You could reach the third floor, where my faculty had its offices, only by climbing to the second floor, crossing the width of the building through a broad corridor where student artwork was displayed, then winding through the tangle of foot traffic up another set of stairs that felt like a fire escape. The annex, where most classes met, could not be reached from the inside of the main building at all. Moving from my office to class, then, demanded that I descend to the second floor, trek across the building to the grand staircase, navigate the congestion at the entrance, walk fifteen steps to a building that was attached but inaccessible, ascend another flight of steps, secure the approval of an annex key lady who could only have learned her craft from the KGB, then climb three more flights of steps and make my way across peeling parquet floors.

If I was assigned a room that had belonged, in the old days, to a senior officer, I had to shout to make myself heard across the cavernous expanse. When my class was scheduled for a cubbyhole designed for an underling and eighteen students showed up, I learned what it meant to live in a city where many people had no indoor plumbing.

The size of my classrooms was the least of my problems. School had begun and I still had no idea how long the semester lasted, when midterms were scheduled or the etiquette of final exams. My request for an academic calendar had been met with a frank "These Americans are so demanding" stare.

What was the grading system, and what was it based on? How much homework was required? Could I put books on reserve in the library? Was there a budget for photocopying? The academic basics eluded me, since the university had no catalog. The course offerings were listed term by term on sheets of paper posted by the entrance. What more did students need, since all their courses were required, leaving no room for electives? Katsev seemed anxious to discuss the theory of journalism, but he showed no interest in teaching me KRSU 101.

When I'd handed out my syllabus, the first printed syllabus my students had ever seen, I'd sensed that my expectations were seriously out of line.

"You want us to write a story every week?" Artyom asked, clearly on behalf of the group.

"Yes, that's the only way you will learn," I responded. Noting the universal displeasure, I quickly added, "But there's no midterm and no final."

The annoyance, obvious in any language, didn't wane.

"Don't you have homework in your other classes?"

"No, we just study and prepare for exams."

I didn't miss the omission of the words *reading* and *attending classes*.

Recovering quickly from their dismay, the students looked for a way out. "Can we write the articles as a group?"

I didn't rebound as quickly. Group work? At a university? I knew that students in the former Soviet Union went through their entire university educations with a single set of "group mates" with whom they attended all their classes. But had they so melded into a single mind that they could no longer work—or even think—as individuals?

"No, not unless you plan to work as a group after you graduate."

I launched my next class with a mock press conference to assess the fourth-year students' interviewing skills. I was the subject, and I'd blithely assumed they'd begin by asking me my name, my age, my professional, even my personal background. Instead:

"Do you work for the opposition press or the government press?"

I was caught off guard, again, and it was only my second class.

The establishment of at least a fictive democracy in the early 1990s meant that the government's monopoly over the media had vanished and that the new opposition media had the freedom to print as many lies and distortions as did the media controlled by, or loyal to, President Akaev. In practical terms, that meant that if *Vecherny Bishkek,* a progovernment daily, reported that the sun rose in the East, *Maya Stalitza,* an opposition paper, would point out the deficiencies of eastern sunrises and the superiority of western ones, and accuse Akaev of having a hand in, or making a profit from, Kyrgyzstan's inferior sunrise position. Predictably, the citizenry was pretty confused about what was really going on in their country. But the concept of an objective press—with balanced news stories that didn't read like editorials—was entirely alien.

"We have no such distinction in America," I tried to explain. "We like to think that we oppose everyone equally."

The fourth-year students were a savvy lot. Three girls had lived or studied in Germany, one in Russia and a fourth in Iran. Tanya was the daughter of a member of the Jogorku Kenesh, the Kyrgyz parliament, and Angela, who I later learned was called "the terrorist" by her classmates, was as in-your-face as her peers were reticent. Their English was so good that my translator didn't have to open her mouth. But they were clearly taken aback by my response.

"But opposing everyone is mean," they complained. "You have to like somebody."

I growled. In my book, the phrase *nice journalist* is an oxymoron.

"It's not my job as a journalist to like anybody. It's my job as a journalist to report the facts so that the people have the information they need."

It was clear that I needed to get more specific.

"What information do people need about your president? Is he honest? Corrupt? Able? Incompetent?"

Akaev had been president since the founding of Kyrgyzstan as an independent nation, and most people still weren't even sure where he lived.

"He's incompetent and corrupt," said Tanya in a tone that brooked no argument. Her father, of course, was a member of the opposition, a leader of the Communist Party. "That's what we need to write."

Kyrgyz opposition journalists had a proclivity for that kind of reporting, denouncing the president as a liar, accusing government officials of theft, usually without a shred of proof. Then they cried, "Persecution of the press!" when they got nailed for libel.

"That's *not* what you need to write," I shot back, emphatically. "Just write the facts and trust the people to come to their own conclusions."

Tanya winced. Trusting the people most decidedly ran counter to local tradition.

"But they won't give us the facts," whined Alexandra, a statuesque blond who dreamed of being an important "news reader," as the Russians called anchors, in Moscow. Whining, I learned, was her specialty.

"You need to dig," I shot back. "That's what I'm going to teach you this year. That's why this course is called American *Practical* Journalism."

I was on a roll, speeding toward a solid, clear explanation of what journalism could be, what it meant to delve for information and write a story without leaving a single fingerprint on your copy. My excitement was palpable; at least it felt that way to me.

Just then, I heard a knock at the door, and a middle-aged woman dressed in a long white coat and a chef's hat strode in and, ignoring me entirely, barked out an order in Russian. My students gathered up their books and began filing out, leaving me in mid-sentence, alone, at my desk.

"Excuse me," I uttered, feeling as pitiful as I sounded. It's tough to feel in charge or maintain classroom decorum when your students vanish without a word.

From the doorway, Angela looked over her shoulder.

"Oh, you didn't understand," she said. "That was the nurse. We must go now for our vaccinations."

"Will you come back?"

"Not until next week," she replied.

Angela slammed the door on her way out.

3

Food for Thought

Bread is bread.
Crumbs are also bread.
—TAJIK PROVERB

Two weeks after my arrival in Bishkek I discovered two avocados teetering precariously atop a stack of fruit boxes in what passed for the "exotic produce" section of our neighborhood bazaar, the corner offering bananas, oranges and lemons. It was my first harbinger that things were looking up—admittedly, an odd choice for a portent, but surely no odder than a four-leaf clover. With avocados, at least, I had the consolation of guacamole if the promise wasn't fulfilled.

We'd been dreaming about avocados, just as we'd been dreaming about roast chickens, juicy steaks, crisp green beans, brownies, corn on the cob, and bacon, lettuce and tomato sandwiches. Even meat loaf had become a gourmet fantasy in a country where we couldn't find anything much to eat beyond bread, rice, fat and greasy cabbage soup.

The first time we'd wandered through Ak-Emmir bazaar, our New York food genes had blazed with optimism. The streets outside were lined with little old babushkas squatting on tiny wooden stools from dawn until dusk to sell eggs with lemon-colored yolks laid by their backyard chickens, apples from neighborhood trees, and a dizzying array of used clothes, rusty drill bits, gladioli, and the occasional toilet seat. The entrance was cluttered with old 1950s-style chrome baby carriages stuffed with freshly baked bread, strangely silent live ducks, hunks of butter the size of bricks and bags of milk that would surely go bad before we got home. As we climbed up a flight of steps, squishing through a decade of discarded produce, beggars blocked our way, small gypsy children with beseeching eyes and wrinkled crones grabbing on to our jackets.

Under the rusted roof of the oversize shed, the market was a dizzying cacophony of music and haggling and the beseeching of five hundred vendors offering everything from bread and condoms to fifty varieties of cheap vodka. Koreans served kimchi, pickled cauliflower and a dozen less distinguishable pickled delights, while the dairy women stirred vats of fermented yak milk, congealed sour cream and what looked like a particularly noxious version of yogurt.

The meat section was a landscape of chunks, lumps and an occasional recognizable body part, usually intestines or a head, hung on meat hooks, displayed on counters or being hacked up by butchers who wielded their hatchets at random. Curious buyers were left to guess what part of the horse, yak, cow, goat or dog their dinner would come from. Chickens no longer producing eggs were summarily plucked and laid on tables in full, unrefrigerated rigor.

The fruit vendors polished and stacked a dozen varieties of apples, creating garish displays to hide the worm-ridden and rotten beneath the perfect. Dozens of vegetable ladies frantically competed to sell the seven approved varieties of produce: tomatoes and cucumbers, potatoes, carrots, onions, cabbage and beets. And the spice ladies haughtily showed off their wares, mounds of red, orange, black and pink powders that, sadly, were all varieties of pepper.

At the nut stands, vendors pressed us to sample five types of pistachios, peanuts, almonds, walnuts, dried apricots and the ubiquitous sunflower seeds, whose shells had hardened into the broken concrete on all the pavements. But woman—at least not this carnivorous woman—cannot live by nuts alone.

From the poultry aisle, we'd chosen two chickens so fresh that they looked like they had been running around outside fifteen minutes earlier. Then we'd bought two *lipiushki*, traditional flat bread with a puffy, dense border. From the vegetable section, I'd picked up a kilo of potatoes, crisp and firm, never touched by a wrapper or a plastic bag, and tomatoes as ripe as the ones I plucked off the vines of my garden back home. And for dessert, heavy cream and a quart of red raspberries the size of walnuts, never sprayed with a single pesticide.

Back home, I squeezed a lemon over one of the chickens, turned the gas gauge to medium—the closest I could come to temperature control—put the potatoes on to boil and invited three friends over for roast chicken and mashed potatoes. An hour later, I stuck a fork into that taut piece of skin that connected the chicken's leg to its anorectic breast, waiting for the succulent juices to run into my spoon. The fork bent. It wasn't a defective piece of flatware; the skin had hardened into a leathery hide that would have protected a rhinoceros from a razor-sharp spear.

An old hand at making do in foreign countries, I don't give up easily in the kitchen. In Peru, I'd taught myself to make mayonnaise without so much as an electric mixer. In Spain, I'd baked my own pita. How hard could it be to figure out how to prepare chicken in Kyrgyzstan?

With my guests almost en route, I dismissed chicken number one as the aberrant bird whose free range had been a marathon course and threw some water into a pot. Then I hacked the other chicken into pieces and put them on to parboil. Having spotted something that looked like paprika in a cabinet, I sent Dennis across the street for sour cream—*smetana*, with the emphasis on the second syllable—so I could whip up a quick chicken paprikash.

The doorbell rang, and we made our guests tea, regaling them with the story of the strange bird. Ten minutes later, I checked my steaming cauldron. Dark brown scum floated atop it, and chicken number two had shriveled, the skin a cruel parody of the brown leather on my couch back home. It looked like I'd be serving mashed potatoes, salad, bread and dessert for dinner.

Except: While the bread and salad were delicious, fresh and full of flavor, the potatoes hadn't mashed so much as dissolved into a gritty powder. The heavy cream—the heaviest available—wouldn't whip, even after 30 minutes in my landlady's admittedly feeble electric mixer.

I crossed chicken, mashed potatoes and cream off the list of ingredients I could use in Kyrgyzstan, and we went out to dinner.

We sampled a Korean restaurant just off campus, a small café with tables on the sidewalk. The kitchen was a corner closet with one wok atop a single-burner gas hot plate, which guaranteed that the four of us would eat our meal serially. The owner spoke only enough English to tell us that none of the dishes we pointed to on the picture menu was available, or that we shouldn't order any of the dishes that we'd requested. We weren't quite sure which, and since none of us spoke either Russian or Korean, we would have had no way of knowing if we were served what we'd ordered in any event.

The next day, my students warned me that even the dishes advertised as beef or chicken were probably dog or horse meat.

Three days later, without even realizing that we were starving, our landlord's son invited us out for shish kebab. Our mouths watered as we drove up to a restaurant on the outskirts of town, a dozen tables set up under trees. The grill by the entrance laced the air with the smoke of searing meat. Saliva began dripping noticeably down my chin.

But when our skewers arrived, the fat-to-meat ratio was seven to one—we actually counted. "You don't like the fat?" our host asked, inviting himself to what we'd left on our plates. "The fat is the best part."

Undeterred, we went through our mental list of "Kyrgyz delicacies" and requested a sampling. The *manty* looked delicious, like oversize Chinese dumplings. But before they were steamed, the meat inside the pastry dough had been topped with a layer of fat. The *plov*, a rice-and-stew combination that was the national dish, was served swimming in—enhanced with, a menu would have read in an upscale dining establishment—gravy made of water and sheep fat. The *lipiushki* and the tomatoes, however, were perfect.

We finally broke down and headed over to the Crostini Restaurant at the Hyatt, the epitome of fine dining in Bishkek. The chandeliers were ruby crystal, and the waiters spoke perfect English. But half the dishes on the menu weren't available, and when my leg of lamb arrived, I couldn't find a single bite of meat underneath the three-quarter-inch layer of sheep-fat gravy—and they served neither *lipiushki* nor tomato salad.

It was back to home cooking.

I bought local salt, which made kosher salt look granular. I purchased beef from four different butchers and fried it, sautéed it, and stewed it. Our teeth just weren't up to the task. I found a German butcher who carried what looked like pork chops, but when I put them on to fry, the pungent odor of urine permeated the apartment. I tried buying lamb in the market. That's when I discovered that Kyrgyz sheep have a strange hump hanging over their buttocks where fat accumulates, and butchers only seemed to handle sliced hump.

I baked a cake, but the pastry flour proved too coarse. I checked out the available fish, but passed when I realized that it all came from lakes and rivers that made Chesapeake Bay seem pristine. The milk and ice cream tasted fine, but it was hard to forget that Kyrgyz dairy products weren't pasteurized and that the World Health Organization found 20 percent of the milk to be swarming with *E. coli,* listeriosis, salmonella, campylobacter, brucella, *Corynebacterum diphtheriae* and *Mycobacterium bovis.*

We tried, combing the most remote corners of every bazaar, every "supermarket," every delicatessen. Day by day, the list of unavailable ingredients grew longer: brown sugar, powdered sugar or molasses, vanilla (extract or beans), unsweetened chocolate and baking powder, green beans, peas, corn, lima beans, zucchini, celery and broccoli, peanut butter, cheddar cheese, maple syrup, corn starch, beef broth, chicken broth, horseradish, dry white wine and dry red wine. Oh, and chicken breasts.

The markets were filled with whole chickens and chicken legs. Legs were rotisseried at tiny stands along the street and served in a dozen guises at restaurants across town. But there was nary a chicken breast to be found.

We were desperate to stroll up to one of the market women and ask, "What happened to the rest of the chickens? Where are the breasts?" But nothing is that easy when you're mute and semiliterate in the local language. Mime is not my forte, but Dennis proved fearless, squawking like a chicken, pointing to my breasts and shrugging what he thought was the universal "why?" to chicken vendors all across Bishkek. Alas, in Russian, chickens don't cluck; they chickchirik. And the all-American questioning shrug conveys a hopelessness that demands no response.

The more we pondered the missing-chicken-part conundrum, the deeper the mystery became, and the more we needed to solve it. When we realized that it wasn't merely that the wings and breasts were missing, but that the legs seemed strangely large for the meager girth of the birds, our need to know turned into nightly practice at chicken pantomime.

Finally, on the third day of our mission, a young Kyrgyz chicken lady in a flowery bathrobe raised her eyebrows at the unfamiliar sounds and awkward gestures of the crazy American. Then her eyes popped open in understanding. Pointed to a chicken leg, she muttered what sounded like *booshleks.* I pulled out my handy Russian-English dictionary and ran down the listings under *beh.* Nothing. I tried the Cyrillic *veh,* thinking I'd confused the consonant. Still no *booshleks.* I handed the dictionary over to our new friend, but she brusquely shook her head. *Booshleks* seemed to have no English equivalent.

I raced home and called Regina, my translator, on the telephone.

"Regina, what does *booshleks* mean?"

Already savvy to the flaws in my Russian pronunciation, she muttered *booshleks* with ten different inflections before declaring, "I don't know this word. I don't think such a word exists. Where did you hear it?"

I almost pleaded for comprehension. "In the market, a woman was trying to help me understand why there aren't any chicken breasts."

Regina guffawed and corrected my pronunciation. "Nogi Busha," she said. "You know, as in the president of the United States. Bush legs."

It seems that during the winter of 1991, when the former USSR was collapsing and food was growing dangerously scarce, Papa Bush, George Sr., sent America's excess chicken parts—all those legs left over from 10 million chicken breast sandwiches—as humanitarian aid to Kyrgyzstan, Russia, Kazakhstan, Uzbekistan, Mongolia, Georgia and the Ukraine. The obscenely oversize poultry gams won the hearts, or at least the stomachs, of consumers accustomed to stringy Soviet-era fowl. Once the emergency was past, aid turned into trade. Bush legs still dominated the market, both for their colossal girth and for their price, half of what locally produced chicken cost.

The second time I bought avocados, they looked and felt perfect. But when I sliced them open, the flesh was brown and entirely rotten. A week earlier, I might have wept. By then, however, I'd moved on to other obsessions, less symbolic portents that we might not spend a year gazing out our windows onto a bleak landscape, eating bread and tomatoes and wondering what we could possibly have been thinking when we decided to move to Kyrgyzstan, of all godforsaken places.

Regina, my translator, was the architect of the turnaround. Hiring a seventeen-year-old translator would be an odd choice in the West, where we equate age with experience and expertise. But in Bishkek, the under-twenty-five crowd was usually more competent than their parents, who'd been raised to a world that no longer existed. What were economists who'd been trained on Das Kapital 101 or administrators schooled in the Dialectics of Bookkeeping supposed to do in a post-Communist world? Suddenly, a Party card and fluency in Bulgarian didn't offer a ticket to a steady job.

No one was ready to admit it, but generational power, or at least the generational balance in knowledge, had been turned upside down, leaving older people clinging desperately to positions for which they were insanely unqualified. Business courses—classes about marketing, accounting, contracts and other techniques of pernicious capitalism—were taught by former collective farm managers and party hacks trained to keep track of how much money their factories lost, rather than by newly minted M.B.A.s just back from Indiana or Nebraska.

Linguists who'd spent years poring over each nuance of the subjunctive but had never had a conversation with an American or a Brit taught English to students who couldn't conjugate a single irregular verb but had picked up the rhythm and tone of the language from music and movies and the mass of foreign missionaries, teachers, consultants and businessmen who'd flooded into the former Soviet Union the instant the Red Curtain parted.

If competence had been the criterion for appointment, Regina, a freshman at the American University of Kyrgyzstan (AUK), an outpost of the U.S. education system in Central Asia, would have been appointed dean of the English faculty at a local university, or at least a senior professor.

Regina's English was the byproduct of the zeal of Australian, American and Canadian missionaries who'd come to preach the Word to those raised by the godless Communists. They'd had so much success that Bishkek alone boasted twenty evangelical churches. One of their converts was Regina's brother, Ruslan.

Ruslan's new friends started hanging around Regina's family's apartment when she was twelve years old. They didn't succeeded in converting her, but they taught her English, and not just of the biblical variety. By the time she was old enough to apply for a high school exchange program in the United States, Regina was tossing off *gee whizzes* and *you betchas* like an American mall rat. After a year in a high school in Texas, she couldn't learn much about English from the teachers in Kyrgyzstan.

We turned our psychic Kyrgyz corner when Regina's mother, Emiliya, invited us to dinner, a culinary event we contemplated with considerable trepidation. Much to the family's surprise, Emiliya's sisters, Dolores and Venara—stalwart communists who'd long refused to meet Americans—had agreed to join us. The prospect of an evening reminiscing old animosities with true-blue Soviets was irresistible, no matter the cuisine.

"So what did you think of the various Soviet premiers?" I asked, scooping up another fork full of *paramyachi*, a traditional Tatar empanada. The Tatars were yet another ethnic group I'd never heard of until I landed in Bishkek. Eastern Europeans descended from Attila "the Great Hun," as they call him, their kingdom had been annexed to Russia by Ivan the Terrible in 1552. Tatars like Regina's parents, however, still retained their religion, Islam, their language, national identity and ability to cook. I'd taken an immediate liking to them the minute Emiliya trotted out a stack of *paramyachi*.

Having been warned that Dolores and Venara might be touchy, I tried hard to stay on neutral ground even as I probed their political sentiments. "Did people like Khrushchev? What did they think of Andropov, of Gorbachev?"

Venera almost spit out Emiliya's homemade raspberry jam at the mere mention of that final name.

"Brezhnev was the best," she said, her eyes almost tearing. "Those were the days. You see us now, we don't have anything. Oh, but back then, life was good. We didn't earn much money, but it was enough. You could fly to Moscow for ten dollars; you could live on your pension and still have something left over.

"But Gorbachev, he ruined everything."

Dolores turned the tables on me, also striving for tact. "What did Americans think of Gorbachev?" I didn't yet understand that virtually everyone in Kyrgyzstan believed that Gorbachev had destroyed their paradise—and that most were convinced he'd done so on purpose, at the behest of the U.S. government. In other words, I didn't know enough to be delicate.

"Americans loved him. We thought he was bringing you freedom. We celebrated when communism fell."

Her face contorted, her anger obvious. "How could you celebrate our misery?"

"We thought you wanted freedom."

"Freedom? Who told you we wanted freedom? And what kind of freedom is this? The freedom to starve?"

"But weren't you afraid during communism? Afraid to speak freely, to disagree with the government?"

"Of course not." Venara bristled, as if the questions were ridiculous. "And we had plenty of money. We all had jobs. We could afford everything."

By then I was too caught up in the conversation to consider discretion. "But surely before, under Stalin, you must have been afraid. I mean, Stalin murdered millions of people and sent millions more to Siberia."

Dolores seized the reins. "It is true that Stalin did some bad things, but Stalin saved us from the Nazis. He modernized our country. He was like a father to us."

Regina didn't remember communism, and like most young people I'd met, she wasn't sure what she thought about "the old system," or the new. Oh, in theory, she thought democracy was a great idea. In practice, she wasn't convinced that it would work in Kyrgyzstan.

"All I know is that life was better before," she told me when I pressed her after my dinner with her family. "Now the factories are all closed, my father is without a job. Maybe democracy works in the United States, but that doesn't mean it will work here. It's not our tradition. We need a strong leader to reopen the factories."

I shook my head. "But they can't reopen factories that weren't making a profit, and none of the factories was making a profit," I suggested. I forgot that while Regina understood the word *profit*, it wasn't part of her daily vocabulary, the type of automatic calculation an American would make.

"I don't care about profit. People need work, so they must reopen the factories."

If Regina was our guide into Soviet Kyrgyzstan, and our distraction from persistent, gnawing hunger, nineteen-year-old Samarbek Ashym uulu led us into what we thought of as the more mysterious, exotic realm, the life of traditional Kyrgyz.

We met Samarbek one afternoon, when we stopped by the "traditional Kyrgyz village" that a local restaurant had created for tourists. No tourists were in sight, even on a sunny Saturday afternoon. But dozens of Kyrgyz from villages—"black Kyrgyz" in local

parlance, to distinguish them from the "white" or urbanized Kyrgyz—had set up their yurts, the felt tents their ancestors had called
home during their centuries as nomadic herders.

Nestled in a parklike setting adjacent to an outdoor café and an
indoor disco, Dasmia was a living museum. Schoolchildren were
brought on outings to learn about their heritage, families celebrated
holidays and anniversaries in yurts decorated with carpets produced
for the fiftieth anniversary of the Kyrgyz Soviet Socialist Republic,
fox pelts, eagle skulls and a riot of colorful Kyrgyz carpets and pil-
lows, all of which were also for sale.

"Hello, can I tell you about the Kyrgyz people?" asked a young
man lounging outside a yurt. That's not precisely what came out,
but that's the gist of what I extracted from his mangled English.
The next thing we knew, Samarbek had invited us inside our first
yurt, made by his mother, Meiman, a skill she had learned, of
course, from her mother—although both had lived in concrete
houses built after the Soviets rounded up the herders and forced
them onto collective farms.

Collectivization had been a brutal dislocation for the Kyrgyz, a
collection of nomadic Turkic clans that had migrated to Central
Asia from Siberia in the tenth century. Although their homeland
had been annexed by Russia in the mid-nineteenth century, czarist
control had been distant and light. But even when a swarm of
Russian immigrants followed in the wake of the Russian army,
fencing in traditional pastures and pushing the Kyrgyz aside,
Kirghizia did not rebel. Only during the harsh, famine-plagued
years of World War I, when the herders were drafted into the Russian army by the thousands, did the Kyrgyz rise up. And they kept
their rebellion through the early years of the Bolshevik revolution,
until the Kremlin decreed that all independent nomads must
become farmers controlled by Moscow.

Born in the south of the country, on a collective farm called
Soviet, Samarbek was the youngest son of one of the senior Soviet
administrators and had spent his first years in a life of comfortable
certainty: He'd start school, join the Oktobronok (the October
Brothers), graduate to the Young Pioneers and wind up in the
Communist Youth. By studying hard, working for the collective

good, and joining the Communist Party, like his father before him, he'd rise to a position of prominence in his community.

Everything had been reassuringly preordained, the real opiate of the masses in Soviet times.

Then, when he was ten years old, between volleyball practice and school, Samarbek sensed that something was awry. The adults around him grew grim and distracted. The Soviet flag was supplanted by a red-and-yellow flag adorned with the skylight of a yurt, the symbol of the new nation of Kyrgyzstan. Then, ominously, the president announced that Kyrgyzstan would follow the capitalist road.

"We'd been taught that working for the common good was honorable, and that behaving like capitalists was shameful," Samarbek explained as his mother hovered and served tea. "People had to start selling their possessions to find the money to eat. They had to do business, to bargain in the market. For people raised in the Soviet Union, that was a humiliation.

"I didn't understand much, but I remember quite clearly that I was afraid, afraid that in the new world of competition, I would be at the bottom, that I would become a slave. That's what we thought capitalism meant: you were either a boss or a slave. A slave was someone who sold his work to another in order to survive.

"We were accustomed to a world in which everyone was equal. I couldn't imagine how I could become a boss, so I was sure that I would be a slave."

I'd been waiting for some sign of Kyrgyz pride, a burst of excitement at Kyrgyz liberation from the yoke of the Russian bear. Hearing none, I probed. "But weren't you also happy to be free of the Russians?"

"Why would we be happy?" Samarbek asked. "We love the Russians. They're our big brothers. The Russians civilized us."

By the time he entered high school, Samarbek had decided that the only way to avoid slavery was to master capitalism. He'd transferred his parents' obsession with Moscow to an obsession with Washington, holed himself in the library to study English, and read everything he could find about the United States and capitalism, a meager selection since few libraries had been updated since the fall of socialism. By the age of seventeen, he'd honed his life's goal: he would become Kyrgyzstan's first Bill Gates.

Samarbek still had a pretty long way to go. His English was fragmentary, he didn't know what a contract was, and the Department of Economics and Business Administration where he studied didn't have a single faculty member who'd ever set foot inside a capitalist country. The youngest of seven children in a culture that coddles the last in line, Samarbek, tall and slim, with dark hair, sharp Asian eyes and the cockiness of a high school wrestler, was nonetheless brimming with confidence.

"I've already made some difficult transitions," he blustered. "When I entered in school, in Soviet times, the only Russian I knew were the words for fish, sheep, cow and horse. I had to master Russian vocabulary and grammar, and learn Russian culture.

"I already know some English. I watch American movies and listen to Britney Spears and Ricky Martin. So it shouldn't be that hard."

I smiled, both politely and with the thought of Samarbek bringing modern capitalism to Kyrgyzstan. Then he interrupted my reverie.

"We have to pass through capitalism in order to get to socialism," he continued. "I'm going to help make that happen."

4

9-11: Sympathy for the Devil

Trust in Allah but tie up your camel.
—MUSLIM PROVERB

"You have to admit that Americans brought this on themselves,"
Angela pronounced, leaning back in her chair, a smidgen of satisfaction framed by her long and wild hair.

The first words that had greeted me when I dragged myself into
class on the morning of September 12 had been expressions of
sympathy and concern. With those second phrases, Angela told me
what was really on my students' minds.

"That's what happens when you meddle in everybody else's
business," added Rada, a stunning young woman with classic
porcelain skin and raven black hair. The whole class joined in. The
message was clear: Americans got what they deserved.

That wasn't the lesson I had planned when I'd designed my
course syllabus a month earlier. But on that morning after, few
Americans anywhere on the planet had the luxury of going about
business as usual.

The night before, at the moment when most Americans were absorbing their first gruesome images of the World Trade Center towers crumbling like Tinker Toys, I'd been drinking tea on the patio of the Orient Café on Ala-Too Square across from Lenin. No one was racing down the street in alarm or weeping in grief. The sunset had turned the snow-capped peaks of the Pamir Mountains a soft pink, and scores of couples moseyed around to savor the perfect fall evening. Only in retrospect did the utterly prosaic quality of that landscape of normalcy haunt me. In it, I measured my distance, my isolation, from home.

Just after 10 P.M., Dennis and I raced through the front door to catch the ringing phone. "Turn on the television," an embassy official whispered hoarsely into the receiver. As we watched the faces of New Yorkers contorted in anguish, the cloud of smoke turning Manhattan's morning into the nation's night, we had no way to make sense of the devastation. Sixty of our sixty-two channels broadcast in Russian, Kyrgyz, Chinese, German or Korean, none of which we understood. Finally, we landed on Fox News, the one American channel on our cable, our only lifeline as we watched America's naiveté crumble along with the heart of lower Manhattan.

The embassy had instructed us to remain in our houses, but the next morning I'd refused to comply, a meaningless display of bravado that had felt like an important blow for New York mettle when I'd staggered out of my apartment thirty minutes earlier. As soon as I set foot onto the utter normalcy of Soviet Street, however, every nerve in my body ached for the refuge of Fox, my only link with home in a world gone mad.

Exhausted from the all-night vigil, I sighed at Rada and Angela's cool certainty, then steeled myself to play teacher.

"Be specific," I instructed. "What 'meddling' are you talking about?"

They all shouted at once: Vietnam, Bosnia, Serbia, Haiti, Somalia, Iraq.

Their knowledge of history, well beyond what American teenagers could call upon, was cold comfort. Could they really not see any difference between Vietnam, which I thought of as old-style American imperialism, and Bosnia, Clinton's postmodern

brand? I wondered why they hadn't added the country I suspected
was really on their minds, the Soviet Union.

"Let me ask you something." I interrupted the litany pouring
forth. "If Uzbekistan invaded Kyrgyzstan in order to annex the
Kyrgyz part of the Fergana Valley, what would you want the United
States to do?"

That lush valley had been split between the Kyrgyz, Uzbek and
Tajik Soviet Socialist Republics by Stalin, no slouch in wielding
"divide and conquer" as a strategy. The Uzbeks considered the
entire valley theirs, a claim seconded by the thousands of Uzbeks
who'd suddenly found themselves on the wrong side of the border
when the Soviet Union was dissolved.

"You must defend us," they said automatically, their voices lacking
any hint of irony or suggestion that they sensed the contradiction.

"But we can't," I responded. "That would be meddling."

"Oh, no, it would be different if the Uzbeks invaded. You
wouldn't be meddling. You would be defending us."

The next hour, I needed a break and took refuge in my syllabus,
which had me scheduled to work on my third-year students' interviewing skills.

"Take thirty minutes to conduct man-on-the-street interviews
about popular reactions to what happened yesterday in New York,"
I told them. "Then come back and we'll talk about what you
learn."

"We're supposed to walk up to strangers?"

"Yes."

They looked totally flabbergasted, but were too obedient to dis-
obey my command.

When they returned, I chalked a list of the most common
responses on the blackboard:

1. *The United States has only itself to blame. If it stayed out of
   everyone else's business, this would not have happened.*

2. *If the United States had not opposed the Russian war in Afghanistan, this wouldn't have happened.*

3. *Terrorists have been blowing people up for years and nobody cared.*

Now that it is the United States that has been hit, the entire world suddenly has to declare war on terrorists.

4. *The United States is scapegoating Osama bin Laden. It has no proof that he is guilty but will go after him anyway.*

5. *The United States is looking for an excuse to test its latest generation of weapons.*

6. *This is the beginning of World War III, a war between Muslims and Christians.*

I'd heard the first argument all morning, so moved on to the second, the sorest of points in a country where so many had fought and died to keep Afghanistan under the Kremlin's thumb. Historical what-ifs are a tricky business, since you can never gauge what will happen to a sweater when you begin pulling out a single thread. The U.S. government had certainly funneled a full measure of weapons to the anti-Soviet Afghan fighters. But would they have lost without that materiel?

"Certainly," my students argued. "The Soviet Army was very powerful. They could only have been defeated with American help."

Given the determination of the Mujahideen and the declining Russian popular appetite for foreign adventurism, I wasn't so sure.

"More to the point, would the disaster of September 11 have occurred if the Soviets hadn't invaded Afghanistan in the first place?" I asked.

My students shied away from the implications of that question.

The second and third reactions were near-universal among older interviewees who still looked to Moscow for guidance. The fourth echoed the voice of youth.

"Let's say, for the sake of argument, that bin Laden *is* guilty," I said. "What would it take to prove that to you? What evidence would you believe?"

The classroom turned quiet; at least my students were thinking.

Angela was the first to speak: "Any evidence the U.S. government showed us could be a lie."

"That's true," I conceded. "Would you believe the evidence if President Putin blessed it with the Kremlin Seal of Approval?"

Their English wasn't good enough for that sort of slang, so I became my own translator. "What if President Putin revealed that he'd seen the evidence and was convinced. Would that be enough for you?"

Several students, all Russians, shook their heads, a vigorous, and telling, yes. The remainder admitted that short of a public confession—"and not if bin Laden is in some U.S. prison," one boy hastily added—no evidence could convince them of bin Laden's guilt. It wasn't that they liked bin Laden; they knew little about the Saudi multimillionaire. They were just suspicious by nature, especially of Western pronouncements.

"So, help me understand why you think there's going to be a third world war." I skipped to the final proposition. "Do you believe that all Muslims hate the United States that much?"

"Muslims don't have any choice but to hate the United States, since the United States is always attacking Muslims," was the response of a quiet girl who'd never before uttered a syllable in class.

"Is that true? Where have we attacked Muslims?"

"I don't know. But that's what people say."

"But in Bosnia and Somalia, we were supporting Muslims," I continued. "And in the war against Iraq after the invasion of Kuwait, we were supporting Muslims who were attacked by other Muslims."

Passing off the information as if it were irrelevant because the equation had already been calculated, a Kyrgyz boy interjected, "The Taliban say this is a jihad."

"Well, is it? And should it be, just because the Taliban promulgated that decree?"

SILENCE.

"My brother thinks it has to be a jihad because the United States wants to kill all the Muslims," said a girl in the back of the room.

"Why does he think that?"

"I'm not sure."

"Do you think he is right?"

"I don't know. I'm confused."

Angela stepped forward once again, "Yesterday the vice president of Palestine said that Israel won't give back the land the Palestinians want."

"Is that what caused the attack on New York?" I asked.

SILENCE.

"What do you know about Israel and Palestine?"

Angela looked miffed, almost insulted. "They're both in the Middle East, but Palestine is a Muslim country."

"What else? What's the history here? How is it relevant to what happened yesterday?"

No one knew enough about Israel or Palestine to continue that line of dialogue.

"But there is going to be a world war between the Christians and the Muslims," one Kyrgyz boy declaimed loudly, almost proudly. I'd just edited his application for a scholarship to study in the United States.

"Which Christians and which Muslims? Are you talking about yourselves?"

"Well, Muslims here aren't really Muslims like in Afghanistan, so we don't mean us."

So much for Muslim unity.

"But Muslims have to defend other Muslims against attack," said the quietest girl in the class.

"What if the Muslims are in the wrong? What happens then? And what happens when Muslims attack other Muslims?"

"Muslims don't attack other Muslims," she insisted, having ignored the earlier conversation.

"Iran and Iraq? They fought an eight-year war. The Iraqi invasion of Kuwait? Should I go on?"

Confusion etched onto her face, she shook her head no, clearly loath to hear anything that contradicted what she'd been taught.

Tired and cranky, I moved the conversation in a new direction: "Can you explain to me why there's never been a call for a jihad against Russia?"

The room tensed. "Why should Muslims be angry at the Russians? The Russians are our brothers."

"But your brothers have been attacking Islam for decades," I continued, ticking off the list: the suppression of Islam in Central Asia, the invasion of Afghanistan, the war in Chechnya.

"But the Russians are poor," they responded.

"So this isn't about religion, then, but about being rich?"

My second-year students were more sober, or perhaps it was simply that Artyom led the conversation in a more sober direction.

"What do you think the United States will do?" he asked.

"I think we're going to bomb Afghanistan," I admitted.

"Then there will be a very long war because the Soviet Union couldn't defeat the Muslims in Afghanistan," interjected Elina.

"Maybe you're underestimating the Americans because you overestimate how powerful the Soviet Union was," I theorized. As I explored Bishkek, I'd been wondering whether Americans hadn't also overestimated the power of the former Soviet Union. If they couldn't build sidewalks, roads, bridges or tunnels that lasted even more than a decade, were their missile silos and bombs really all that much better?

Unwilling to confront that possibility, the shattering of a lifelong myth, the students veered off: "This is very bad for Kyrgyzstan."

They'd clearly run through the possibilities overnight.

"If we support the United States, the Taliban will attack us."

"Refugees will stream across our borders, and we already have too little to feed ourselves."

"The Taliban and other Muslim fundamentalists will hide among the refugees."

A new edge had crept into their voices, an abrupt shift in tone from resentment and envy to thinly veiled fear. They'd just begun to catch a glimmer of modern possibility—of the freedom of books and music and travel. That panoply of choices terrified them, but not as much as the specter of being confined to *borqas* and beards by a medieval theocracy.

Frankly, I couldn't take the prospect of thousands of Afghan refugees streaming into Kyrgyzstan very seriously. The Afghan border lay just 350 miles south of Bishkek, but those 350 miles included the Tajik border, which was guarded by Russian troops, and some of the highest mountain passes in the world.

And nothing about local life and history suggested that scores of Kyrgyz were rushing to answer the call to jihad, leaving the country in danger of becoming Afghanistan II. Afghanistan, after all, hadn't begun to move into the modern world until King Amanullah took the throne in 1919—and within a decade he was fleeing Kabul in his Rolls-Royce after igniting the wrath of the nation's warlords by inviting his queen to unveil. By the time any other Afghan monarch summoned up the courage to try out another such noble—or ignoble, depending on your perspective—experiment in Kyrgyzstan, socialism had long replaced Islam as the warp binding daily life. Factory work had superseded herding and peasants from remote villages were flying to Moscow for education and technical training. When the Soviet Union fragmented, some Kyrgyz families, ironically mostly former Communist Party faithful, had taken refuge in Islam. But, as my students were quick to say, "We're not that kind of Muslims." Few villagers seemed to be beseeching Allah to bring in the Taliban.

The Kyrgyz, however, had been hearing a steady drumbeat of warnings about the danger of the Afghan war spilling over their frontier, and the kidnappings of the last three summers had created near-panic. Ahmed Shah Massoud, the preeminent leader of the Afghan Northern Alliance—assassinated two days before the strike on the World Trade Center—had cautioned, "The Taliban will not stop at seizing the whole of Afghanistan! Their plans include creating a supranational radical and extremist Islamic state." And the leader of the World Islamic Union, Osama bin Laden, had announced, publicly and proudly, "We are going to purge Tajikistan, and then the whole of Central Asia."

But too much about the threat rang hollow. President Akaev echoed Massoud's warning, but mostly when he was bidding for more aid from the Americans. When he played to the local audience, he minimized the dangers: The rebels aren't interested in Kyrgyzstan, they're just crossing our territory to get to Uzbekistan, the country they really hate. Or he accused the troublemakers of being drug smugglers using Islam as a cover, hoping, it seemed, to discredit them among Muslim activists. (That line struck a chord with a population still naive about political rhetoric. But how many drug smugglers call attention to themselves by kidnapping foreigners and issuing press releases to the international media?)

The other Central Asian presidents followed the same script, mining the prospect of a fundamentalist takeover as a convenient political ploy rather than facing it as a serious threat. The president of Tajikistan ran for cover, pretending that he had no knowledge of insurgents operating from bases in his country, despite the fact that the Islamic Movement of Uzbekistan had maintained a strong hold there for years. And Islam Karimov of Uzbekistan used the IMU raids and kidnappings as yet another pretext for rounding up members of the opposition.

The IMU wasn't entirely a paper tiger. Founded in 1996 by Tajik dissidents and Uzbek fundamentalists who'd fled Uzbekistan when President Karimov had ordered draconian measures to suppress Muslim extremism, the group seemed intent on overthrowing the Uzbek government and had set off a series of car bombs outside the parliament, leaving sixteen dead and sparking a vicious counter-jihad. But if they hadn't been toting Kalashnikov assault rifles, grenade launchers, handguns and knives during their summer raids into Kyrgyzstan, the fundamentalists' jihad might have been a Muslim parody of the Keystone Kops. Rebels trained in terrorist camps in Tajikistan, Chechnya and Afghanistan had stumbled across 18,000-foot passes without hiking shoes or sleeping bags, gotten lost in the unmapped Alar Range, headed toward Uzbekistan, then veered back deep into Kyrgyz territory. During ransom negotiations, they couldn't make up their minds whether to ask for $1 million per hostage, $500,000, or $2 million, the figure changing by the hour.

When I arrived home after class, my e-mail in-box flashed open with forty-seven messages from friends in New York, none of them providing the reassurance I craved, all instead displacing their anxiety onto me.

"Are you all right over there in Bishkek?"

"This is not the time to be so close to Afghanistan. Get out before the bombing begins."

"Come home on the next plane!"

"We're hearing rumors about fundamentalist terrorists all over the -stans. COME BACK!!! NOW!!!"

Most of my friends hadn't been thrilled five months earlier when I'd divulged my plan to spend a year teaching in Central Asia, imagining I was moving into harm's way. Since my parents had lived in Jerusalem for almost twenty years, I'd understood their angst all too well. For years, at each report of an attack on an Israeli taxi, a bomb in the market, the impending bombing of Iraq, I'd placed frantic phone calls to Israel:

"Are you okay?"

"Please come home."

"What do you mean, you just picked up your gas mask?"

My mother's reply was unrelentingly calm: "Do I call you every time I hear about a murder in the United States?"

But as I'd prepared to leave home that summer, it had never occurred to me to silence my friends that pithily, because it never occurred to me I might have as much cause for worry as did they. After all, I was the one leaving the haven of America, which to all the innocent conventional thinking of Americans, including me, meant I was the one tempting fate and courting chaos. They were staying put, which meant that they were staying snug.

Bishkek still felt reassuringly peaceful, at least by comparison with New York. Our Kyrgyz friends had been immensely kind. Our family—our American family—had been attacked and maimed, so they, a people who honored family, overwhelmed us with condolences. Neighbors called to inquire about our loved ones. Strangers stopped us on the street and shook their heads, sympathy overcoming the language barrier. In the market, an Iranian woman wearing a tightly bound headscarf approached us. "People say to me, 'Aren't you glad?' " she said in broken English. "Who could be glad about such a thing?"

When we picked up two rugs we had purchased from an Uzbek family, the mother handed them to us in a paper-wrapped package, neatly tied with string, and added, in a tone we found entirely unreadable, "Don't worry, there are no bombs inside."

Late that night, we strolled down Soviet Street to our local Internet café, one of dozens that dotted a city in which people had the notion, but rarely a convenient means, of connecting with the wider world. It was packed, twenty or thirty locals trying to get through to friends and family in America.

An inveterate snoop, I eavesdropped as best I could on the conversation between a couple drinking beer at the next table, also waiting their turn on-line.

"The mullah condemned what the hijackers did as a violation of Shari'ah," declared the young woman, dressed in cheap Chinese knockoffs of the latest fashion from the cover of Russian *Elle*. She seemed comforted by that pronouncement, a confirmation of the gentle form of Islam to which most Kyrgyz were raised, perhaps even a validation of her affection for things American, evident in her KISS T-shirt and the videotape of *Moulin Rouge* hanging out of her faux Prada bag.

The young man at her side disagreed athletically: "The U.S. deserves what happened yesterday." Dressed in the international uniform of young men—jeans and a polo shirt—he gave no indication that he was a devout Muslim. No beard, no Muslim cap.

I sensed pride in his voice as he talked about the attack, mimicking the boom of the crash of the airplanes and the roaring implosion of the towers, and perhaps I was imagining it, but it read as, "Wow, young Muslims, guys like me, stood up to the giant and toppled one of its icons. We're not all puny, irrelevant bumpkins after all."

He wasn't glorying in the death and destruction, nor was his tone vicious or bloodthirsty. But it was as if for just a moment, his connection to the hijackers, no matter how bogus or ephemeral, had made him feel less like a kid from some forgotten fragment of a has-been empire that history had condemned. Incapable of identifying with those confident, strutting Americans, so secure in their comfort, the young man from a world suffused with frustration and hopelessness had lit up like a winner.

5

Doctors and Development Junkies, Missionaries and M.B.A.s

Don't sweep where you don't put your yurt.
—KYRGYZ PROVERB

In Bishkek, English wasn't merely a language, a prosaic form of communication to facilitate expressions of love, exchange of confidences or commercial transactions. It was an obsession imbued with the mystery and glamour of California beaches, overstuffed refrigerators and cars zooming down highways as smooth as the bronze of Lenin's cheeks. The capital's 850,000 residents all seemed to have concluded, simultaneously, that the reason the empire had crumbled, the economy collapsed and the government turned to corruption was because they didn't speak the right second language.

The over-forty crowd didn't try to make sense out of the ABCs, but even in remote villages, elementary-school children carefully practiced their Latin-alphabet letters in old-fashioned primers. Students at the nation's sixty-three universities labored to decipher the difference between *the* and *an*. And the minute a foreigner opened his mouth in a store, at a bus stop or in the Opera House, twenty young people pounced to plead, almost forlornly, for a quick practice session.

Kiosks and telephone poles were plastered with advertisements for English courses and teachers, and families scraped together every spare *som* they could summon to pay for classes, private lessons, special tutoring and correspondence courses that would magically guarantee their children's futures. Those like Samarbek, who scraped and scraped and still couldn't amass a very hefty pile of *soms*, went to the weekly English classes at Radio OK.

Since we didn't spend a lot of time surfing the airwaves in Russian and Kyrgyz, we'd never heard of Radio OK until the night Samarbek stopped by on his way to class. Curious about how his mother tongue was taught, Dennis decided to go along.

"Do you mind if my American friend enters?" Samarbek asked politely after they'd trudged up five flights of steps in yet another building with an elevator that no longer functioned. Peering through the door into a room bare of anything beyond a circle of thirty chairs, Dennis didn't miss the annoyance that flitted across the face of a thirty-something Australian woman wearing a skirt that almost brushed the floor. The cross around her neck shone brightly, even in the dim light.

"Why does he want to attend?" She clearly was not pleased.

Dennis strode into the room, intent on defusing the tension. "I just thought it would be interesting, and I'm happy to help out," he explained, taking a seat at the circle between an ambulance driver, a physician and a university student, all of whom immediately wanted to talk. By then, he didn't need to hear their questions to reel off the answers: Dennis, from New York, my wife teaches at KRSU and, yes, I like Kyrgyzstan very much.

The class opened with the predictable routine of going around the circle for introductions and explanations ("My name is Nazgul and I want to learn English because English is the world language, or English is very good language, or with English I make much money"). Even the next step seemed pro forma: "Students, could you please fill out these forms?" said the Australian with the unctuous tone Sister Olive had affected when Dennis was in grade school.

The forms, however, had no spaces for names or addresses. They inquired, instead, "What sort of music do you like?" "What radio stations do you listen to regularly?" "What can we do to improve Radio OK?"

Dennis was taken aback for a moment. Then his MBA kicked in: the students had handed over 150 *som* for an English course and wound up as focus groups for Radio OK marketers. He tensed at the blatant deception, and the teacher didn't disappoint his mounting sense that something was amiss.

"Tonight we have a special treat," she cooed. "Wes Hendrickson from Radio OK is coming in to address your questions."

What happened to the advertised English lessons? Dennis wondered. And what would anyone want to ask the head of Radio OK? The students, however, obediently broke up into the groups to prepare queries for their august visitor.

"I have some questions," Dennis told his companions, the doctor, the ambulance driver and the university student. "I want to know how he justifies using the class time of paying students for a marketing survey. And I want to know who owns Radio OK, who pays for it and what religious affiliation it has."

You don't have to spend much time in Central Asia before your missionary radar kicks in, and Dennis's had been bleeping wildly.

"You can't ask questions like that," Dennis's group mates told him. "We don't ask questions like that in Kyrgyzstan."

Dennis would not be denied.

"If you're going to learn English, you need to learn how we behave like an American," he told them, writing his questions down in block letters so that Azamat Aitoktor uulu, the university student who was his group's "presenter," could read them out loud. With his wide face and ruddy cheeks, Azamat could have graced a tourism COME TO KYRGYZSTAN poster. He beamed with open exuberance and, at the age of twenty-five, was as playful and curious as a child, winning over foreigners with warm curiosity and a bottomless willingness to endure the ritual humiliation that is learning English. Asking pushy questions, however, ran decidedly against his Kyrgyz grain.

By the time it was Azamat's turn to ask questions, Dennis had discovered that Wes Hendrickson was married to Karen, that they had adopted Kyrgyz children and that they liked Kyrgyzstan very much. But Hendrickson was less forthcoming about his financial backers in the States, and Azamat refused to press the issue of the marketing survey or of Hendrickson's religious affiliation. Since Radio OK, FM 100.5, blared Christian music almost twenty-four hours a day, sandwiched in between news, on-air English lessons and the advice of Christian psychologist James Dobson of Focus on the Family, however, Dennis didn't really need to hear his answer.

"This is a front for Christian missionaries," he declaimed to Azamat, who'd followed Dennis and Samarbek to the bus stop.

That wasn't a paranoid delusion. In much of Central Asia, missionaries live undercover, using code words like *tentmakers* for what they're doing and talking about working for "the Company." That habit made sense in Afghanistan during the Taliban, where missionary activity was illegal. But missionary activity was legal in Kyrgyzstan, and Christian churches operated openly and undisturbed.

"But I'm a Muslim," said Azamat, confused by what Dennis was telling him. "My father was a Muslim, and his father before him."

As we came to discover, Azamat, the twenty-five-year-old son of a Kyrgyz family so traditional that he made Samarbek look like a hip-hop kid from Watts, rarely talked about himself without embedding his present in his family's, or his people's, past.

We'd been hanging out with urban Kyrgyz for more than a month, young professionals so assimilated by the Russians that many no longer spoke their national tongue. Even Samarbek was absolutely bilingual and bicultural, one foot planted in the past, the other on his imaginary path to corporate superstardom.

Azamat, however, was a member of the most traditional of the Kyrgyz clans. When he'd shown promise as an artist and the State had proposed sending him to Moscow to study, his parents had refused. "They worried that I would forget my Kyrgyz language and traditions," he said, regret and resignation mingling in his voice. "They didn't want me to stop being Kyrgyz, so I didn't get to study art."

The prospect of being converted, then, alarmed him. "I didn't sign up for English to become Christian," he said. "Do you think they'll refund me my money?"

Radio OK wasn't the only outfit set up to give Azamat's traditions a makeover. They didn't call their endeavors "makeovers," of course, preferring less candid descriptions like "introducing the Lord to the heathens," "modernizing the infrastructure" or "bringing the education system into line with the Western world."

But Bible Mission International was in the midst of a campaign to "plant" 300 evangelical churches in Central Asia, and Christian singer Rebecca St. James had flown to Bishkek to popularize the Gospel. Missouri Synod Lutheran missionaries ran a mobile medical/dental trailer, a church in Bishkek, and a series of village "preaching stations." And vowing to turn the Silk Road into a "highway for the Gospel," evangelicals filled an outdoor stadium in Almaty with the faithful to hear Norm Nelson of Life at Its Best radio ministry in Orange, California, witness their holy spirit.

Behind the missionaries had come gender-equity specialists and poverty alleviators, water engineers, prison reformers, housing experts, democracy consultants, teachers, lawyers, physicians, nurses, drug control analysts, defense advisers, mine clearers, small business mentors, accountants and AIDS experts sent by universities, the World Bank, the Asian Development Bank, the American Bar Association, the United Nations, the European Union, the governments of the United States, Germany, Turkey, Great Britain, Switzerland and a raft of NGOs—nongovernmental organizations. Danish consultants were renovating village water systems, Helvetas found an American lawyer to teach property rights to farmers whose collectives had just been carved up into private parcels, Pragma was teaching accounting to small bakeries, an international journalists' group was unionizing reporters, the Swiss government was developing a market for yak meat, and the Soros Foundation was sponsoring everything from photography projects to libraries.

They constituted the fastest-growing industry in a country without any other industry.

In places like Kyrgyzstan, the number of NGO-owned Toyota
Land Cruisers and international consultants pulling in monthly
salaries ten times the local average annual income exist in inverse
proportion to the per capita income of those they come to help.
Kyrgyzstan was loaded with Land Cruisers, and the consultancy
business was so hot, until September 11, that Hyatt Regency had
opened a hotel with a nightly room rate of $90 per person.

Each international NGO spawned at least two dozen spinoffs,
local NGOs allegedly heeding the call for the creation of the sort
of strong popular institutions so vital to democracy. Most were
shams, one- or two-man shows set up by Kyrgyz professionals
who'd mastered the art of grant writing as a way to pay their bills
and earn passage to international conferences. What did groups like
the Center for Social Research, the Center of Young Researchers,
the Center on Socio-Political Problems and Connections with
NGOs actually *do*? Not much beyond feeding the insatiable international NGO appetite for information that would allow them to
secure yet more funding.

Dozens of women's groups had burgeoned overnight, promising
to help women cross the "digital divide," train village women in
small business development, organize art exhibits, sponsor conferences and keep all the other women's NGOs organized. But how
were these NGOs liberating women? Seemingly, one by one. Give
every woman her own NGO and international funding, and liberation will surely follow.

Yet Kyrgyzstan was still a mess, sliding backward, the rate of
decline accelerating as the bottom loomed alarmingly near. The
locals had mastered the lingo of the international aid community,
and they could banter about civil society, democracy, participatory
decision making, private property, competition, diversity and capitalism. But ten years after independence, most people just didn't—
or wouldn't—get it, a lesson I re-learned almost daily in the
classroom.

"What is a 'scoop'?" a student asked me one morning, having
heard the term in an American movie the night before.

"It's when one reporter finds out important information before
all the others," I said.

"But that's mean," my students complained. "It's not good for the society."

"No, it's not mean, it's competition," I explained, launching into an exegesis on the American belief that competition is actually good for society because it pushes individuals to try harder, to be more clever, and companies to cut prices. I hadn't yet discovered that their Economics 101 class included no survey of the basics of capitalism.

"It's not cutthroat," I was quick to add, defining that phrase. "We call it *friendly* competition."

Impossible, the students exclaimed. "Competition can't be friendly. It destroys social relationships."

Blinded by their enthusiasm for bringing the Word to Kyrgyzstan, foreign advisers ignored one simple truth: "Getting it," as they dubbed their goal, demanded at least a tacit admission that the old ways hadn't really been so great or, God forbid, that they had been a disaster, a leap the Kyrgyz remained loath to make.

The first time I witnessed the resistance flare was during a lecture by the assistant director of AIDS programs for Kyrgyzstan, who'd crisply asserted that there were only 150 cases of HIV in the country. The number felt absurdly low to me, so I pressed the woman during the time set aside for questions.

"How many people have been tested?" I asked.

"150,000," she responded. You don't have to be much of an AIDS expert to know that you can't derive a national infection rate by counting up the number of positives that turn up in 3 percent of the population.

"Have you done projections based on those 150,000?" I asked. In the late 1980s, I'd reported on AIDS for the *Miami Herald* and knew that the Centers for Disease Control in Atlanta, the World Health Organization and the Joint United Nations Programme on HIV/AIDS all had pretty sophisticated software that allowed them to derive those projections based on scientific sampling.

"No."

"Then how do you know that you have only 150 cases?"

"We can't do projections, it's too expensive," she finally admitted. I could see her back arching, defensiveness kicking in.

"But won't the U.N., the U.S. or W.H.O. help you out with the
technology?"

"We don't believe in projections," she objected, clearly citing a
passage out of a Soviet textbook in epidemiology that she'd read
twenty years earlier at the university.

There was no point in arguing, of course. What could I say?
Most of what you were taught is outdated, and much of the rest
was just wrong? But I ratcheted my optimism for Kyrgyzstan down
a notch in what became almost my weekly exercise.

A well-meaning and stubborn lot, Bishkek's expatriates were not
easily discouraged. But they labored with at least one hand tied
behind their backs, a disability imposed on them by the firmest pre-
cept of overseas aid work: Thou Shalt Not Be Disrespectful of Local
Traditions.

I admit that I couldn't quite grasp that bit of political correct-
ness. If local traditions were so worthy of respect, what did the Kyrgyz need us for? How do you teach Capitalism 101 without
stepping on traditions like collective farms and nomadic herding?
Were all those gender-equity specialists bent on abolishing customs
like bride-kidnapping, arranged marriages, veils and wife-beating
really interested in honoring tradition? And by what semantic
sleight-of-hand could a missionary justify the conversion of Muslims to Christianity as "respecting local tradition?"

No one was fooled, of course. No one could be—the conflicts
between the traditional and the new were too keen, too public,
especially when it came to religion.

I heard about them almost daily from Dinara, one of my second-
year students. Smart, insightful and focused, Dinara had returned
from a year studying in Iowa intent on reshaping her nation by
becoming its first female president. But she'd become involved with
a local evangelical church, and its members—ever so mindful of
their responsibility to respect local traditions—were pressuring her
not only to tell her conservative Muslim family that she'd converted but to try to bring them into the fold.

"Keeping your conversion a secret is disavowing God," they advised her. "And missionizing is part of Christianity."

Trying to balance the unbalanceable, Dinara was torn apart as she tried desperately to be a good Christian yet to remain respectful of her family. "My father is very religious," she told me, explaining that he'd taken refuge in Islam after the collapse of communism. "I don't know anything about Islam. I'm not a Muslim. But my family will reject me if I tell them that I've become Christian."

Nonetheless, the foreign NGOs all tried, or at least pretended, to walk that strange tightrope act. And no group tried harder than Habitat for Humanity, a progressive charity carefully crafted to remove any whiff of paternalism from their work. Tanya and Eric Weaver were running the international organization's first foray into Central Asia, and the city of Bishkek had given them an old apple orchard with enough land for nineteen homes and a small park. Eric and his local board had designed small brick houses with water, electricity, bathrooms and gardens, and the board and staff had carefully chosen their first owners. Contracts had been signed for the first ten units and supplies ordered. Construction had just begun.

Except: Half of Bishkek wanted a Habitat house since they were the only folks around who didn't demand cash up front for dwellings, mortgages being another one of those insidious capitalist schemes that had never existed in Kyrgyzstan. In their excitement, the chosen few hadn't paid close attention to the fine print: Future owners must help pay for their house with "sweat equity," by laying bricks, hauling cement, wiring electricity and sawing boards for all ten houses. Only when all the units were completed would they be assigned their own homes.

Habitat's way, a model of Christian brotherhood, turned out to be as culturally insensitive as an old-fashioned charity's. For seventy years, the Kyrgyz had been forced to work collectively. Once solidarity with their neighbors ceased to be de rigueur, the gloves had come off. Cooperation? They were over it. At the first mention of the word *cooperate*, which they heard as "collective labor," seven decades of suppressed individualism ran riot. Day after day, the new

owners went about the business of building houses—the ones they
had chosen for themselves. When they were caught by the on-site
Kyrgyz overseer and instructed to join the collective roofing group,
the cement group or the siding group, they complied—until the
overseer departed and they could get back to their real work, on
their own homes.

By November, chilled that my students seemed more inclined to
emulate reporters for the old *Pravda* by rewriting press releases than
to become hard-nosed journalists, I was ready to admit that I hadn't
traveled halfway across the planet to respect Kyrgyz or Russian culture, neither of which had produced anything approaching a decent
newspaper. I was there as an agent of change, and I was fed up with
everyone else's pretense that they weren't.

Call me crazy. Call me a pushy, disrespectful New Yorker. But I
just didn't see why the poverty alleviators couldn't say, straight out,
what they acknowledged over drinks at the Navigator restaurant:
Look, your traditions are no longer working for you; if you don't
change, you're going to starve. Couldn't the missionaries admit
what was in their hearts, that they were trying to convert Muslims
because they believed that they needed Jesus, or would go to hell
without his grace? Or the gender specialists share the horror they
felt when a battered woman cried, "It's just our tradition?"

Lying, after all, by omission or commission, is the ultimate form
of disrespect.

For three months, I tried to be a good, tolerant, "when in
Rome" liberal, but I was pushed over the edge by a weekend trip
to Lake Issyk-Kul, the jewel on which Kyrgyzstan was pinning its
bid to become an international tourist mecca, "the Switzerland of
Central Asia." I never discovered who came up with that catch-
phrase, but it had spread like wildfire—across Kyrgyzstan.

"It makes sense," my students argued. "We are friendly people,
our mountains are very beautiful and Issyk-Kul is the most beautiful lake in the world." It was recited like a well-practiced mantra.

Renting a car and driver, Dennis and I had taken off for
Karakol, a small city at the eastern end of the lake, hard on the Chi-

nese border. We sighed in relief when we left behind the air of Bishkek, permanently polluted by coal and wood smoke, heading, we imagined, for a pristine alpine lake surrounded by snow-capped mountains. But the road was rutted and potholed, so the 225-mile drive took seven full hours, without a clean bathroom in sight. The scenery was . . . well, nice, but more like the rolling hills of eastern Montana than Lake Märjelensee at the end of Switzerland's Aletsch glacier, or Lake Tahoe, for that matter. Karakol's only hotel with reliable hot water was at the far end of town across from an abandoned factory, a good thirty minutes from the beach, and the beach's only changing rooms were the bushes, which also served as the local toilet. The sand was littered with trash. Oh, and the rusting hulk of an old factory peeked out of the water.

Clearly no one had bothered to inform the Kyrgyz that beneath Switzerland's peaks sat four-star hotels, cozy country restaurants, highways smooth as glass, bathrooms and pristine sand.

Rather, the tourism fantasy was fed with easy advice by foreign NGOs, foreign embassies, the World Bank and the United Nations. Just ease your visa restrictions, consultants urged the Kyrgyz government. Abolish the old system of demanding "letters of invitation," which were really just bribes to local travel agencies, and make it cheaper and less complicated for visitors to come, they counseled, as if a reduction in visa fee from $65 to $30 was going to fill the empty Hyatt Hotel.

Increase publicity, others suggested. Go to international tourism fairs, buy advertisements. Show Europeans and Americans what you have to offer and you'll open the floodgates. The Kyrgyz Ministry of Tourism dutifully followed that suggestion, pushing Kyrgyzstan as the new, undiscovered Nepal and a skiers' paradise. European travel agents arrived to check out the offerings and went home wondering, Where are the Buddhist monks, the temples? Adding, Oh, and by the way, Western skiers are used to lifts to take them to the top of mountains. Most don't want to walk or spend $1,000 on a helicopter.

Nonetheless, the World Bank kept funding a dozen tourism-related initiatives, the Germans lent Akaev a special tourism adviser, the U.S. government chipped in by giving money to small travel companies to find their business footing, and the World Tourism Organization offered technical assistance. Helvetas organized shepherds and farmers willing to open their yurts to foreigners willing to forego toilets, showers, privacy and edible food for a peek at the "nomadic" life, and local businesses borrowed money from friends to revamp hotels, open small restaurants or buy jeeps—and waited for the hordes to descend.

In 2000, 24,000 visitors arrived from countries beyond the borders of the former Soviet Union, and tourism contributed a mere $4 million to the gross domestic product. A half million Kazakhs and Russians still went to Lake Issyk-Kul, a popular resort during Soviet times. But they weren't the type of tourists who could pull Kyrgyzstan out of its economic hole; on average, they spent just $7 a day. Helvetas bragged that the number of tourists going to yurt camps was skyrocketing, up 300 percent in a single year. However, that was an increase from 55 to 170—hardly the stuff of an economic boom.

Shortly after our trip to Issyk-Kul, I had lunch with an embassy official, and we spent an hour over greasy omelets trading jokes about the search for usable bathrooms between Bishkek and Karakol.

"Why do all the foreigners encourage Kyrgyzstan in this fantasy about being a new Switzerland?" I asked. "The roads are terrible, most of the hotels are uninhabitable, the waters are polluted and the country is in the middle of nowhere."

He sighed in embarrassment.

"The situation here is so bad, but people are so nice that no one wants to hurt their feelings."

part TWO

6

Unembeddable

He who doesn't risk never gets to drink champagne.
—RUSSIAN PROVERB

So, your editor calls you on the phone and says, "I want you to go to Afghanistan to write about the situation of women." What do you do first?

It was November 1, and I was trying to pick up the pace of my courses. We'd practiced basic interviewing techniques, plotted reporting strategies and discussed the problems of achieving balance. Now I wanted to give them a big, complex story and see if they could lay out how to approach it.

"First I would say no, I won't go," said Angela.

That was definitely not the conversation I'd envisioned.

"She's right," added Rada. "It's too dangerous for a woman to go to Afghanistan."

I couldn't resist, my usual problem, which is why we were two weeks behind schedule.

"And it's not too dangerous for a man?"

Leaving that thought dangling, I moved on to the topic of the day.

"Okay, let's say that you *do* agree to write the story. How do you begin?"

I could see from the easy expressions on their faces that they didn't find my question a challenge.

"You go to Afghanistan," said Angela, seeming bored.

"How do you get there?" I threw it back quickly.

Angela looked disgusted, clearly thinking, Why is this woman wasting our time?

"I go to a travel agency and ask them," she replied blithely.

"But the airlines aren't flying to Afghanistan in the middle of a war," I said, surprised that she'd walked into that trap. "And most of the borders are closed. So how do you figure out how to get there? And aren't there things you should do before you leave?"

Dennis and I had been struggling to unravel those same problems for a week, so I knew that my students were about to receive a lesson in the tangle of niggling practical details and grand concepts that is the challenge of reporting.

Afghanistan had not been on our itinerary when we'd mapped our Silk Road fantasy. After all, the Taliban were about as friendly to women as they'd been to the two ancient Buddha statues they'd blown up. The closest we'd thought to come was to Iran, a trip we were arranging for winter break. But I'd just watched a riveting BBC documentary built around interviews with the former dean of the law school in Mazar-i-Sharif and one of Afghanistan's leading writers, both female, and realized that in the dozens of articles I'd read about the plight of Afghan women, I'd rarely seen the faces or heard the voices of Afghani professional women. It had been all too easy, then, to avoid the reality that many of the women being stoned for adultery or whipped for taking off their gloves in public were an awful lot like me.

I dashed off a quick note to *Elle* magazine and started fantasizing my reporting.

Women's magazines are better known for sending their reporters to fashion shows in Paris or Milan than to war zones. But, unable to distinguish between satin and shantung, I'm not the usual women's magazine reporter, and Laurie Abraham, executive editor of *Elle*, was certainly not the usual women's magazine editor.

"Do it," she responded before I had even figured out how to find the women I'd decided to interview.

Since they lived somewhere in Northern Alliance territory near the Tajik or Uzbek borders, I assumed that the easiest route to their doorsteps was through Uzbekistan. I'd catch the one-hour flight to Tashkent, take a taxi south to Termiz on the border and be ready to cross the Bridge of Friendship over the Mau Daryl River before nightfall.

The Uzbeks, however, had mined the bridge, their main crossing to Afghanistan, and were refusing to allow even the United Nations and the World Food Program to use it.

Scratch Plan A.

If you want to know how reporters find their way into dicey situations, check the datelines of their stories. On their way in or out, they give you a clue: For this war, they seemed to be moving in through Tajikistan. Even better, I thought. Dushanbe, the capital, was a short hop by air and less than 120 miles from the Afghan border.

Dennis, who can cajole even the most obstreperous bureaucrat into succumbing to his will, made a list of tasks: Afghan visas, Tajik visas, Afghan and Tajik press credentials, flights to Dushanbe, a car to the border. Two sets, of course. Years earlier, we'd agreed that Dennis would always go with me into danger; he'd rather be captured by my side than be left home to worry.

Meanwhile, I began mapping out my reporting. Some journalists prefer to land cold in foreign cities and move around by instinct, precisely what Angela blithely assumed she would do. That's not my style. I download maps, read everything I can manage and arrive with twenty phone numbers, a dozen addresses and five appointments. So I went on the Internet and started my search for the women I'd seen in the documentary.

Fifty e-mails later, I was stuck. No one at any of the scores of Afghan refugee or women's groups around the world had heard of the female dean of the law school in Mazar or the writer I'd seen interviewed.

I'd been using my Afghanistan planning as an ongoing teaching technique, leading my students through what I hoped would become their future professional process. When I found myself stuck, without any interviewees, I turned to them for suggestions.

"So, what do you do next?" I asked.

"Give up?" Rada suggested, still clearly appalled at my plan.

For five hours each evening, I searched the Internet, sending e-mails to every Afghan woman, every Afghan refugee and every Afghan refugee support group for which I could find an address. I didn't need the women I'd seen in the documentary. I just needed professional women who hadn't fled the Taliban, and somebody, somewhere, would be able to lead me to them. I found dozens living in Pakistan, Germany, England and the United States, but no one could lead me inside Afghanistan's borders. Then I received a note from California. "Do you know Munavara Gul Ahmad in Bishkek? She should be able to help you."

It figured.

Three days later, I walked down the street—just four blocks from my apartment—to the office of the Afghan school and refugee center run by Munavara and her husband. A mechanical engineer trained in Kabul and Leningrad, Munavara was one of seven sisters, all graduates of the University in Kabul in fields from engineering to veterinary medicine. She greeted me from an old beaten-up couch, where she was curled up in jeans and a sweater, looking like she belonged on the Upper West Side of Manhattan.

"You can't understand what it was like when the Taliban came unless you understand what our lives were like before," she told me as her assistant served us tea and she pulled out an album of fading family photographs: her girlhood bedroom, a guitar leaning against one wall; a group of friends at a Kabul café; her mother in a bathing suit at a resort in the Hindu Kush mountain range. They were Muslims, more or less, but pretty much the way that most American mainstream Protestants are Christians.

"For the Taliban, the liberation of women was symbolic of communism, and of everything they thought was the worst, the most decadent and evil about the West," Munavara said. The communist government in Kabul had outlawed bride price, bride selling, the murder of brides who weren't virgins. They'd vigorously promoted the education of girls and the professional training of women. "The struggle for the soul of the nation, for the fundamentalist soul, was fought over the position of women."

Suddenly, the clock was turned back to an era Munavara barely remembered. One morning she arrived at work to discover that men and women were to be segregated into separate offices—more than a small inconvenience, since their jobs were inextricably intertwined. Then the new government began floating the notion of creating a Ministry of Women, where all female state employees would work. When it was pointed out to them that there was no way to mix physicians from the Health Ministry with engineers from public works, they gave up on the idea, and Munavara breathed a sigh of relief.

"I thought, okay, they're not that crazy," Munavara continued her tale. "Little did I know."

A few weeks later, the Taliban announced that, beginning the following morning, women would be required to wear *borqas* outside their homes. "We didn't have any *borqas* in our house," Munavara explained with a bitter laugh. "We didn't know anyone who had a *borqa*. They belonged to women in the villages. Not even my grandmother had worn one."

So much for working, Munavara and her sisters concluded, assuming, correctly, that the next step would be a total ban on female employment. They stayed home, at first reading, relaxing and watching Indian movies on television. Then television was banned and women were required to cover their windows with heavy drapes to shield them from public view.

After a friend was beaten because she'd exposed an inch of her wrist while checking the size of underwear in the market, Munavara decided that she'd had enough and headed up to Mazar-i-Sharif, beyond Taliban control, to find a way out of the country. "I vowed that I would not go back so long as there was a Muslim government in Afghanistan," Munavara said, her body taut with fury.

Her husband, also a refugee, put his hand on her knee, then turned to me and cracked, "Islam, Islam, Islam. You can't imagine how tired we are of hearing about Islam."

Kyrgyzstan offered Munavara and two of her sisters visas, and over the next two years, they saved enough money to bring their father and their sisters out, one by one. Before the last, Nadia, a high school principal, could join them with her husband and children, the Taliban took Mazar. Nadia, her husband and two daughters were left stranded in Kabul.

"Why don't you go interview her?" Munavara suggested.

I trudged home through the snow, which had fallen early that year, and found Dennis in the kitchen steaming cauliflower, a rare find.

"Hi, honey, do you want to go to Kabul?" I asked as I hung my coat on the hook by the door. Dennis shot back the same look of astonishment and secret delight that flared across Ricky Ricardo's face each time Lucy proposed another harebrained scheme.

The United States had been bombing Afghanistan for four weeks, and the *New York Times* was predicting that the Taliban wouldn't surrender any time soon. Obsessed with finding Nadia, I blithely decided that by the time we made our arrangements to leave, the Taliban would have surrendered the Afghan capital, a classic act of journalistic faith that the story would not elude me.

We booked a flight to Dushanbe for November 28, sent optimistic e-mails to our Tajik contacts to meet us at the airport and began the long process of securing visas.

Years earlier, I'd spent three frustrating months lobbying for a visa to Cuba in order to interview the sole survivor of an American community founded on the Isle of Pines in the early twentieth century. An elderly woman displaced by the meanderings of history, Edith Sundstrom had invited me for Christmas. Beginning in September, I placed a weekly "So, is anything new?" call to the Cuban Interests Section in Washington, D.C., trying to be ever-so-friendly and chatty in order not to alienate the press attaché I was bugging. But when December 20 arrived and my visa approval didn't, I exploded, "What is wrong with those people in Havana?"

The press attaché had proved to be a patient man. "Ah, my dear, don't take it personally. Cubans learned bureaucracy from the masters, the Russians."

He wasn't kidding.

The Tajiks, who'd seized on the invasion of the international press corps as a way to pay off their $1.23 billion external debt, demanded letters of invitation from local travel agencies, letters of introduction and $75 for each entry visa, another $100 for reentry rights, $100 for a Tajik press card that was entirely unnecessary since I planned no interviews on their soil, and another $200 for permission to cross the 120 miles from Dushanbe to the Afghan border.

The Afghans, who'd spent only one decade in Russian bureaucratic training, required a single visa. But the embassies abroad, controlled by the Northern Alliance although they controlled little of Afghanistan, couldn't decide how much to charge. In New York, they wanted $30 and a letter from *Elle*, but I wasn't in New York. The embassy in Dushanbe had picked up the Tajik predilection for extortion, demanding $200. And the consul in Kazakhstan, told Dennis that the fee was $150 the world over, but we'd have to send a driver to Almaty with our passports.

Conquering the bureaucracy became Dennis's full-time job.

The day before my departure, I asked my students to help me prepare interview questions and frame enough of my story that I wouldn't lose time once I hit the ground. Oddly, they seemed too distracted to concentrate.

Finally, Artyom, the Russian boy with the ponytail, a deadly serious young man with a sly twinkle in his eye, interrupted: "Aren't you afraid?" The query stopped me cold. Frazzled by visa hassles, ever-changing airline schedules and my reporting, I'd forgotten entirely to consider fear.

I've flown in and out of some pretty lame airports—a cow pasture in rural Ecuador comes to mind—and on some pretty bad airlines. One flight I took from Cochabamba to La Paz, Bolivia, turned around ten minutes after departure to pick up a passenger who'd arrived late. Cubana de Aviación insisted on serving dinner despite the absence of tray tables, forcing me to balance my drink on my lap as we passed through the Bermuda Triangle.

But flying in the former Soviet Union is in its own category, a demented sort of time travel complete with vintage aircraft, carbon-paper passenger manifests and food produced before the invention of expiration dates. The process of moving millions of people from point A to point B was designed to produce full employment, not passenger service, which meant that dozens of clerks, all doing jobs that could easily have been handled by one person, had to pretend that they were important.

At the airport in Bishkek, then, we had to wait in line to pay a $10-per-person exit tax before we could queue up for check-in. When we finally made it to the counter, a woman examined our tickets, then handed them to another woman, who stamped them before turning them over to a third, who checked our names off the list so that a fourth could issue our boarding passes. Since we'd been foolish enough to check our backpack, stuffed with sleeping bags, notebooks and enough packaged food for a week, we had to wait in line for the man who would weigh and inspect it, making him the fifth human being involved in our preflight check-in.

Passport control involved two immigration officers, who examined our documents page by page, and a third whose job description was to usher us through a metal detector that was not plugged in. The final hurdle was a surly woman who rifled through our tickets, passports, exit stamp receipts, visas and luggage checks, in case her predecessors had missed a twenty-dollar bill hidden inside a fold.

By the time we made it to the hard plastic chairs in the cold waiting room filled with standing ashtrays planted directly underneath NO SMOKING signs in at least three languages, we were exhausted. There was nowhere to buy a magazine or make a phone call, and the bathrooms hadn't seen a clean mop since Khrushchev denounced Stalin.

After thirty minutes, our forty fellow passengers abruptly raced to the double door as if a silent whistle had been blown, each maneuvering toward the front with the special Russian three-finger push executed with slight pressure to the waist of the obstruction-ist. The American concept of personal space shattered under the relentless assault, my attempt to resist it with overt shoving sum-marily denounced with hostile glares, much headshaking and "Rude *Americanski*" muttered in Russian, Tajik and Kyrgyz.

I assumed all of this meant the plane was ready to board, but ten minutes passed before the door—never both doors, part of what we thought of as the Soviet "one-door" policy—opened and our fellow passengers squeezed through and onto a bus, a strange interlude since the airplane was parked no more than forty paces away.

When the driver offloaded us by the aircraft, an old Tupelov that Aeroflot had surely abandoned a decade earlier, the passengers gathered around the stairway until the stewardess rechecked all the tickets. Finally we crouched into the tight cabin, pushing past passengers trying to squeeze copious amounts of luggage into a small hold at the back of the plane. Predictably, the baggage overflowed onto the center corridor, which turned out to be handy since the plane was overbooked and the extra passengers left to stand in the aisle at least found perches on the suitcases.

Forget amenities like air vents and reading lights—but at least both engines functioned. Oh, and the stewardesses served drinks, a choice of water or tea, hard candy optional to sweeten your drink.

We held our breath during the ninety-minute flight to Dushanbe, the Pamir Mountains just inches below us; at least that's how it felt. But when we touched down, nobody moved.

"Isn't this Dushanbe?" I asked Dennis. "Why are all the other passengers staying seated?"

Finally, the pilot emerged from the cockpit. Only when he had disembarked did the passengers gather their belongings and deplane. No hurry, no pushing. All the frenzy had been expended on boarding.

Landing was a relief—until we were diverted from the path to the terminal and escorted into an incongruous patio framed by rusting semi containers and fenced in by corrugated metal. A guard who spoke not a word of English and could barely utter a phrase in Russian motioned us with his gun into a container, the back end of an old semi, that had been converted into an office. There, he indicated, we should wait. For what, we had no way of knowing.

"He must be going to get an immigration officer to give us visas, or an interpreter," I said to Dennis, who was already beginning to shiver. An electric space heater sat under the desk of the official, but he seemed disinclined to share the warmth.

"I've been waiting five hours for an interpreter," said a man sitting forlornly on a stool in the corner. A physician from India, he'd been brusquely denied boarding onto his plane to New Delhi and had been languishing there ever since. His request for access to a telephone so he could call the Indian Embassy had been denied, although he'd been granted the right to use a filthy toilet in the courtyard.

An hour passed. Still no one appeared.

"Have you tried offering the guards money?" I asked our Indian companion.

"Twenty minutes after they put me here," he responded. "They weren't interested."

With that, I became officially worried. Who ever heard of a Tajik official who wasn't interested in a bribe?

A second hour passed. It would be dark in a third, and I didn't relish spending the night on the filthy floor of a container located somewhere on the back lot of the Dushanbe airport while six Tajik guards giggled at the foreigners. I thought about trying to sweet-talk our guards into letting me call the U.S. Embassy but gave up when I remembered that the U.S. Embassy in Tajikistan had been moved to Almaty, Kazakhstan, three years earlier.

I read all the posters and notices on the walls, or at least the few I could get through without a Tajik dictionary: duty rosters, customs regulations, nothing about the imprisonment of foreigners, thankfully. Trying to translate them was better than trying to figure out if I was more terrified of being shipped back to Bishkek, missing out on Afghanistan, or being locked up for a week inside the filthy lime green walls of a seven-by-twenty-foot semi home.

My only solace was the fantasy that either our Tajik travel agent or a representative from the Soros Foundation, who'd both promised to meet us, was outside lobbying for our release. In Central Asia, there's nothing more comforting than the thought of savvy insiders wheedling visas out of immigration by calling their second cousins, who lived next door to the officials' brothers-in-law.

It was lucky for me I had no way of knowing that both of my "saviors" had been informed that the Kyrgyzstan Airlines flight we'd arrived on did not exist.

Just as night began to fall, a new official entered, our passports, invitation letters and visa applications in his hand, and requested $75 each. We couldn't hand him the money fast enough. In fact, we handed him extra, motioning him to keep the change, lest he have a change of heart and disappear into the emptiness of what was beginning to feel like the airport of a ghost town.

Our guard abruptly motioned to a heavy metal gate, unchained the lock and led us on to the street. Distracted by sheer relief, we'd walked almost a block before we realized that we had no idea where we were or how to find a taxi, that we didn't speak Tajik and had not a single Tajik *somoni* in our pockets.

"Hey, wait a minute," Dennis yelled to the guard still rechaining the gate. Somehow, because the adrenaline of desperation moves you beyond your normal limits, Dennis managed to communicate the dilemma to our former captor, who led us to a taxi, negotiated a price to the Tajikistan Hotel and paid the driver out of his own pocket.

The afternoon suddenly made even less sense.

A member of the Best Eastern Hotel chain, the Tajikistan Hotel was a modern eight-story structure overlooking Dushanbe's Central Park. The telephones worked, the televisions broadcast BBC, the Business Center had installed extra computers and fax machines, and the lobby was packed with journalists wandering aimlessly— skeletal, nervous, alternately silent and muttering. Most had just returned from weeks on the ever-shifting Afghan front lines, sleeping in cold huts, catching rides with military vehicles down bombed-out roads, eating whatever food they could scrounge from Afghans willing to sell an egg to a foreigner for $1 each, a loaf of bread for $2. Some had been inside the prison at Mazar-i-Sharif four days earlier when a riot by Taliban and Arab prisoners was quelled in a sea of blood. Others had watched as their friends were murdered by bandits on the Jalalabad road. A small group had arrived the night before carrying the body of a Swedish camera- man who'd been killed in Taloqan by a fifteen-year-old bandit.

Gaunt and haggard, they couldn't summon up an ounce of

bravado or braggadocio. That would come later. They ate, they slept, and they warned: "Don't even think about going to Kabul overland."

We'd left home without a plan for getting from Dushanbe to Kabul, knowing, at that point, that we'd have to fly by the seat of our pants. Our goal had been to take a car to the border, cross the Pyanj River on the old Russian ferry that was pulled across by a tractor on shore, then catch a ride with a military convoy straight into Kabul. But that route, we learned, was insanely dangerous: Hazara bandits were menacing anyone moving around the north, and desperate Arab holdouts fleeing Kunduz were roaming the hills. The roads were pocked with bomb craters, and the 1.5-mile Salang Tunnel through the Hindu Kush Mountains had been destroyed, forcing those desperate to get to Kabul to scramble on foot over boulders and bomb rubble in complete darkness. The fifteen-mile journey would take more than a week, they said, a week of begging for food and paying $25 a night, when we were lucky, for an infested blanket on the floor of a mud-brick hut.

Chilled by those reports and the grim visages of their colleagues, dozens of journalists and aid workers were stranded in Dushanbe looking for a way in. The Tajik military had posted a sign-up sheet for helicopter flights down, at $1,500 a person, but no chopper had appeared in six days.

We pushed the problem of transportation to the back of our minds and concentrated on "regularizing" our presence in Tajikistan. We needed press passes, permissions to leave the capital and visas to reenter before we could face the challenge of finding our way south.

The next afternoon, we'd just gotten back to our hotel, our papers in hand, when a young Russian guy sidled up to me in the lobby and asked quietly, "So, do you still want to go to Kabul?"

According to the press badge dangling around his neck, Anton was a fixer for the BBC, a guy whose job entailed everything from bribing officials hesitant to issue visas and arranging transportation to places without roads to leading reporters through arcane bureaucracies. He'd been flown down from Moscow to help the BBC's reporters move in, around and out of Dushanbe.

"You bet," I responded, as if we'd been best friends for twenty years. I'd never seen Anton before, but I was hardly going to tell him he was confusing me with some other middle-aged American reporter he'd promised to help.

"Don't tell anyone else," he whispered, jerking his head toward the swarm of journalists pacing the lobby, yelling into their cellphones to calm down editors back in New York, London, Tokyo or Beijing who wanted to know why the hell they were still in Tajikistan. "Be in the lobby at 6 A.M. and bring $600 each, U.S. cash up front."

The first streaks of lavender and pink were painting the sky when our taxi took off through the empty Dushanbe streets, following the cab into which Anton had loaded his two BBC correspondents. When we reached the airport, he guided us in through a back gate onto the tarmac. There, alone in the middle of the field, sat an aging IL-76 Russian cargo plane—the workhorse craft that had transported 90 percent of Russia's troops to Afghanistan during their calamitous attempt to subdue the country—surrounded by a Russian crew, Russian troops and Yuri Brazhnikov, Russian deputy minister for emergency and calamity relief. They were flying a portable hospital to Kabul, the first humanitarian flight scheduled to land at Bagram Airport, thirty miles north of the capital. Anton had made an "arrangement" with the pilot to carry a few extra passengers.

As dawn turned to day, with no immigration or customs formalities, we hefted ourselves into the cargo bay, flung our backpacks on top of haphazardly stacked boxes of medical supplies and flew south. In the IL-76, the navigator sits in a glass compartment in the nose of the plane at the bottom of a two-story cockpit, the sky seeming open beneath his feet. Squeezing in beside him, we gazed breathlessly at our first view of Afghanistan's parched mountains and isolated villages as we flew for two hours over the Hindu Kush into Bagram, too thrilled at the scenery, and our coup, to consider what would happen when we landed.

We didn't circle the airport or wait our turn to touch down.

There were no other aircraft headed for Bagram, no ground crew to guide us to our gate. The old Soviet airbase was deserted, the windows in what had once been the main terminal shattered, the doors padlocked. Bombed-out planes littered the runways, the only traffic dried grass wafting lazily along them. The only sound disturbing the utter stillness was the explosion of mines at the far end of the runway, being set off by American and British troops.

"Don't set foot off the tarmac," the Russians cautioned us as we jumped out of the plane. "There are land mines everywhere."

There was no government in Afghanistan, but even if there had been, our expensive Afghan visas probably still would have been irrelevant to the squat Afghan soldier with a machine gun who ran the airbase like a one-man fiefdom. A local BBC fixer haggled with him to open the gate and let us out, with no success. The British and American military, recently ensconced at the Bagram, sent in their flaks to negotiate, according the nasty little sergeant the respect due Kofi Annan. But reason and negotiation failed. Cold cash did not. Forty dollars later, the gates parted and we hitched a ride into Kabul with the BBC.

I don't think anyone spoke during the trip, as our van negotiated a powdery road pitted by the bomblets from cluster bombs past deserted villages and charred gun emplacements. The landscape was studded with ruined mobile rocket launchers, their rockets fused into the tubes; abandoned tanks; and gnarled bridges suspended in the air midriver. The few trees still standing had been denuded of all but their uppermost branches for firewood. Eight-year-old boys toted rocket-propelled grenade launchers, the playthings of children born and raised in a world of war.

Ninety minutes later, we forced our way past a line of taxis and shills, around Toyota Land Cruisers with United Nations logos and guards with pointed rifles, up the steep drive to the front entrance of the InterContinental Hotel. Below us lay an arid plain ringed by mountains whose slopes were dotted with tiered stone and mud houses: Kabul. The scene before us was as chaotic as a neurotic's nightmare.

In the foreground, camera crews loaded and unloaded satellite dishes and state-of-the art digital cameras; below us, donkeys pulled homemade wooden wagons up the street. Every Kabulite who spoke twenty-five words of English circled the front door or paced the lobby, looking for work as a translator, at $60 a day, one-fifth of the annual per capita income.

"I can take you to many bodies of dead Arabs."

"I know many good womens you can interview."

"You need good driver not afraid of Talibs and drives very fast?"

Not a single one of the InterContinental rooms, all pockmarked with bullets, was available, not even on the floors without any water. Najib, a fourth-year medical student and our newly minted translator, whisked us off into downtown. Just as darkness fell, we checked into the Spinzar Hotel. Two weeks earlier, it had been the Taliban guest house.

The manager, a young man whose closely trimmed beard suggested that he was hip to the new program, took our $90, the price of a week-long stay, and turned us over to the sixty-three-year-old chamberman—or whatever you call the elderly men who replaced the chambermaids when the Taliban outlawed paid female labor.

"The water sometimes can be hot if you tell me in advance," he explained, in English. An old-school type, he seemed comfortable with foreigners, perhaps because he'd worked for five years helping the Taliban entertain "foreign dignitaries."

Our room was huge and utterly basic: two sagging twin beds, a nightstand devoid of the laminated wood that had once covered it, a sink and a closet sans hangers.

"Might I get a lamp?" I asked, noting that the ceiling light was too dim to read by.

Ten minutes later, he knocked on our door and produced an aging floor lamp, its pink shade drooping and discolored. Shaking his head in apology, he stripped the wires at the end of the old black cloth cord and stuck them directly into the socket.

"It's something old we had stored in the basement," he said, smiling sadly.

Nervous and anxious and somewhat incredulous, we rinsed our faces with cold water, descended the stairs, nodded to the uniformed guards playing with their Kalashnikov assault rifles at the entrance and gingerly walked outside.

As our eyes adjusted to the utter blackness, we caught sight of a cluster of dim lights and faintly flickering candles that suggested that a few eggs were still left to be sold, that canned milk and fresh bread still could be bought in the market down the street. Picking our way around mounds of garbage, we waited for someone to ask us. We'd thought long and hard about what we would say. Canadians, we'd decided. It would be safer to become Canadians. But as it usually happens, the moment mugged us.

"American?" a group of small boys ran up and exclaimed excitedly. A dozen older men watched warily from the sidelines of what swiftly became a circle.

"Americans, yes," we answered, warily.

"Americans good!" one boy shouted.

"New York," Dennis tentatively, motioning a friendly thumbs-up, unaware that thumbs-up wasn't particularly friendly in Afghanistan, where it signals "fuck you."

"New York very good," the boys cheered.

The older men drew closer, and one took Dennis's hand in his and yelled, "America–Afghanistan, New York–Kabul, good, very good."

They offered us tea. They shook our hands, again and again. Then the children followed us back to the Spinzar, where we climbed to our room, wondering who had slept in our beds before us.

7

Braving the *Borqa*

*A man with a frog head is better
than a woman with a golden head.*
—KYRGYZ PROVERB

It wasn't until the third year of the Taliban regime that Nadia Sidiki lost the feeling in her legs.

The first year, the former high school principal was too grateful for the end of five years of civil war—of endemic rapes and looting by thugs in the armies of seven competing warlords—to absorb the reality that her life had been reduced to the walls of her apartment and a blue shroud that even her grandmother had not worn.

The second year, she managed to convince herself that the reign of terror couldn't last, that the power of the semiliterate zealots recruited out of Pakistani religious schools would collapse or that they would ease their relentless war on Kabul's women. After all, how long could the city function with 70 percent of its teachers and almost half of its doctors locked inside their homes?

By the third year, that hope could no longer sustain her. Her two oldest daughters were losing what little knowledge they'd acquired before they were forced out of school in the sixth and third grades. Their eyes glazed over in disbelief when she reminisced about the old days in Kabul—the Kabul before Afghanistan became equated with Islamic fundamentalism—when she and her six sisters studied at the university, laughed with their classmates at cafés, and shopped for the latest European-style clothes in fashionable shops in the new part of the city.

Trapped inside her *borqa* and her apartment, her face grown pasty from years without sunlight, the thirty-two-year-old woman who'd always been the liveliest member of her boisterous family could no longer delude herself that she'd get her old life back any time soon. "It was like a slow poison," she told me, perched on the edge of a table in her living room in suburban Kabul.

I'd appeared on Nadia's doorstep my first morning in Afghanistan carrying letters and photographs from her sister, Munavara, in Bishkek. She'd swept me into her tiny living room—one wall adorned with a curio cabinet stuffed with yellowing plastic toys, college textbooks, postcards of Kabul in the 1980s, the detritus of a dead world—and hadn't stopped talking for three hours.

Her apartment building was eerily reminiscent of her sister's back in Bishkek, cut from the same Soviet cloth, assembled from identical unadorned concrete slabs. But Nadia's courtyard was virtually denuded of trees, and her lightless bathroom held only a toilet. By comparison, Munavara's apartment looked like a cheery townhouse on the outskirts of Los Angeles.

A broad grin lit Nadia's face, her excitement at having contact with the "outside" palpable. But as she choked out her tale, the grin turned shy, as if what she was talking about was too absurd to be real, as if slightly embarrassed by her hunger for confession.

"At first, you don't realize what's happening. You refuse to believe what's happening. By the time you can no longer deny it, it's too late to counteract the venom."

Her feet turned numb, and the lack of feeling soon crept up into her thighs. One morning, she could no longer walk.

Her neighbor Jamila Mujahed never lost her ability to move, but she had nowhere to go. Afghanistan's most prominent broadcast journalist, Jamila had led the news team at Kabul Radio for a decade, reporting on the 1978 coup that gave Afghan leftists control of the country, the Russian invasion a year later, the ensuing war between the Soviet army and the fundamentalist Mujahideen, and the capture of Kabul by Mujahideen militias in April 1992.

By the summer of 1996, dread enveloped her as she reported on the Taliban's steady advance toward the capital. Wherever they established control, women were forced out of work, girls barred from their schools. The voices, even the laughter, of women was banned from the streets.

The last news she delivered was the announcement that the Taliban were poised on the edge of the city. The following morning, Kabul Radio was renamed Radio Shari-ah, and Jamila was no longer welcome at the microphone.

"Once the Taliban arrived, my education, my professionalism, become irrelevant," explained Jamila, second on the long list of interviewees I'd developed before I'd left home. "I was used to staying late at the station working on my stories, drinking tea with my colleagues, male and female. I was a free woman. I was suddenly deprived of my liberty, reduced to some distorted and crazed view of what it means to be a woman so that all women—peasants and intellectuals, writers and workers—would become alike."

Jamila stopped speaking for a moment and pulled back into the light scarf that covered her hair to regain her focus. Her husband, Said Amin Mujahed, reached over to caress her, his steel-gray eyes heavy with grief, and she picked back up again, her voice even and resonant. "I was filled with anger, with rage, but I had no way to express it."

Caught between fury and depression, she couldn't sleep and took out her anger on her husband, a gentle man who stepped down as chairman of the history department at the Academy of Social Sciences when the Taliban Ministry of Culture refused to heed his plea that they preserve the ancient Buddhas in the village of Bamiyan. Sometimes she lost control entirely and lashed out at her children, smacking them for no reason, or no good reason, or meaning to hit them lightly, only to wind up drubbing them until their hysteria broke through her own frenzy.

Finally, like Nadia and an incalculable number of other educated Afghan women, she sought professional help. The physician she consulted, however, concluded he could do nothing to heal her: "The only medicine I can prescribe is the medicine I can't give you. The only cure for your ailments is going back to work."

Nadia's doctor had been equally powerless when she'd limped into his office, leaning heavily on her husband. He assumed right away that her paralysis had no physical origin and counseled her to avoid stress, Nadia says, laughing at the absurdity of giving that advice to a woman living under the Taliban's steadily tightening chokehold of restrictions. All he could offer was the same comfort that dozens of medical personnel at hospitals and underground clinics prescribed for thousands of Kabul's women—tranquilizers, regular doses of Valium, lithium and neurozene.

Men, too, were ground under the heel of the primitive system of justice put in place by the Taliban, who cracked down on the population with as much vigor as Stalin, and with the same consequences for apostasy. Those driven to thievery because their wives could no longer earn any money were led out onto the field of the soccer stadium so that thousands could watch while their hands were cut off. Those whose beards weren't long enough to protrude below a fist clasped at the point of the chin, or whose clothes were deemed inappropriate, were summarily whipped on the street.

Najib, our translator, spent a week in jail because he was caught playing chess. Safiullah Rasheedi was imprisoned for allowing a woman to pull back her *borqa* to examine a dress that caught her eye in his clothing store. The husband of a young Pakistani-trained hairdresser was dragged into the police station for selling cosmetics under the counter. And the week we arrived in the country, the noses and ears of twelve men were sliced off after they'd shaved their beards, a grim reminder of what male Afghans, too, were escaping.

But everything in Afghanistan was about gender. While men were punished for what they did, for defying the authorities, the Taliban's relentless war against Kabul's middle-class women wasn't a simple equation of crime and punishment. There was a desperation to the floggings and the stonings meted out to these women, a hysteria behind the ruthless and uncompromising humiliation that spoke to just how powerless the fundamentalist leadership felt in the new world that had taken root in Kabul, a world inching toward modern values like diversity, tolerance and equality. The sight of an unveiled rosy cheek or a lock of hair tousled by the wind was the most intimate, thus the most potent, symbol of a world they could no longer control, despite Allah's command that they control it. Only by concealing that swath of flesh, by reining in those unruly tresses, could the men in power regain the sense that they were masters of their perversion of a Koranic domain.

When I first arrived in Kabul, like journalists from all the other major continents—from Chinese television and Polish radio, Japanese newspapers, Slovakian monthlies and African newsletters—I hunted down Kabul's women to catalog the indignities they had suffered, as if the horror could be measured by the length of the list, weighed in the goriness of their beatings.

I didn't have to look far: a young woman limped across the courtyard of Jamila's apartment complex. What had happened to her leg? Unaccustomed to navigating the streets in her *borqa*, she'd tripped on the sidewalk during the first months of the Taliban, exposing a bit of flesh, and had been beaten with a studded leather strap by the religious police, the psychopaths from the Ministry for the Protection of Virtue and the Prevention of Vice.

An elderly woman picked through tomatoes in the market, her hands cruelly scarred. Her crime? Washing clothes outside without gloves. A middle-aged lawyer—at least that's what she'd been "before," as Kabul's women politely put it—wandered the streets of Micrayon Six in the suburbs, mumbling and muttering. One afternoon, she and her husband had had an argument on the street and she'd pulled back her *borqa* for an instant because it's impossible to engage in marital discord without flashing your eyes. A religious cop noticed and beat her senseless. She'd just been released from a Pakistani mental hospital, where she'd lived for three years.

Then, one afternoon, over tea—even where there was nothing else, there always was tea—Jamila challenged me to a different calculation: "Imagine what it would be like for you, for American women, if your freedom was suddenly hijacked by maniacs from another century and you were helpless to get it back." Only then did I realize that the true measure of the horror wasn't so much in the sensational stonings or the dramatic beatings as in the mundane reality of living imprisoned by those possibilities: Spending years with your drapes tightly closed lest you run afoul of the religious police. Learning to examine meat in the market or pick up bags of rice without revealing so much as an inch of your wrist. Watching your daughters like a hawk lest they race outside and talk to neighborhood boys. Wondering if you'd slip in the rain in front of the wrong person or tear your *borqa* and not be able to go outside at all. Turning on the radio during the broadcast of the week's stoning at the soccer stadium.

Just as they sought out the most gruesome tales, foreign journalists combed Kabul looking for signs of resistance to this nightmare, and they lionized the teachers who'd set up underground schools in their homes or doctors and nurses who'd hoarded drugs for illicit clinics. But the need to find, and celebrate, dramatic gestures, or even timid rebellion, is a peculiarly Western obsession. Afghanistan's women weren't looking for the heroines among them. For them, survival had been sufficiently heroic.

Kabul looked like a living set for a movie about medieval Europe, except for the dim electricity and the cacophony of blaring radios, honking car horns and backfiring buses that competed with the bleating sheep. There was no government, but the traffic police still tried to create order out of the chaos of yellow Toyota taxis, garishly painted Pakistani trucks overflowing with bulging rice bags, carts built from truck chassis pulled by wrinkled old men, children pushing wheelbarrows cobbled together from buckets and bicycle wheels, buses so full that people perched on the roof or hung out the open doorways, and hundreds of brand-new Toyota Land Cruisers. The mail was being delivered. Clothes stores, metal shops and a hundred repair stalls fixing everything from teapots to new trucks filled the cavities of the disintegrating buildings. And the markets overflowed with fruit and vegetables, grain, jewelry, radios, videotapes, and American jeans, sweatshirts with Harvard logos and

Fairfield County High football jackets, all donated in good faith by folks in Middle America and sold for the right price to the people of Afghanistan.

In the early morning, Kabul redefined *still*. There was literally no one on the streets, as if the actors and crew hadn't yet arrived for their day's shoot. Then, just after dawn, a few brave souls emerged, and within thirty minutes, the city came to life—until 5 P.M., when night fell like a curtain and the crowds evaporated into the descending darkness. Kabul was brown, the brown of endless rows of mud-brick buildings and a fine—almost liquid—dust covering every structure, every road, and the clothes of a throng of people. The only color breaking the monotony was the yellow of a thousand taxicabs and the clear blue of the occasional *borqa*.

At first, it felt impossible to fathom the shape of prewar Kabul—a city of art museums and movie theaters, elegant shops, smart restaurants and female senators back before mile-square swaths of the city were reduced to fields of rubble during the fighting of the early 1990s. There was hardly a wall that wasn't pockmarked with bullet holes. The electricity flickered between faint and blackout. The telephone system was so weak that you couldn't call from one neighborhood to another.

All that remained of the old palace was a bombed-out hulk, the five glass domes cracked, the interior picked clean by the hungry. Scavengers dug through the ruins of old homes and shops looking for scrap metal. Metalworkers beat Coke and Sprite cans into satellite dishes that MoMA would have hung next to their Warhols.

But I caught a glimpse of the privileged world in which Jamila, Nadia and her sister Munavara had been raised when I stopped by the Herat Restaurant for lunch with my photographer. Wherever reporters converge en masse, the Herat arises in one guise or another: the American Colony Hotel in Jerusalem, the Foreign Correspondents Club in Phnom Penh, Cambodia. Ask any taxi driver in Mogadishu or Kuala Lumpur where the journalists hang out, and they'll take you to the local variety of the Herat.

Journalists had saved the Herat from extinction after the departure of the restaurant's steadiest clientele, the "foreign Taliban," as westerners called them, "the Arabs," in local parlance, always enun-

ciated with a tellingly angry spit. The foreign Muslim militants—
the men from Chechnya and Pakistan, Saudi Arabia, Indonesia and
Egypt who saw their own futures in the Taliban's triumph—had
filled the Herat's tables each afternoon to gobble down plates of
rice and *ashok*, kebabs and *mantu*, to drink tea, puff on water pipes
and trade news. After November 12, 2001, journalists followed that
same pattern, served by the same waiters who'd brought food to
bin Laden, ordering off the same English-language menus created
for the old customers whose lingua franca, too, had been English.

The only change was the appearance of the occasional woman,
foreign women like me, given the honorary status of men,

After lunch, I took a break from my interviews, and Dennis and
I pushed through a press of hungry women in filthy *borqas* begging
for food, hanging on our coats for money, pleading in Pashtun or
Dari for relief from reality. We meandered down the tree-lined
streets of Wazir Akbar Khan past an empty children's playground,
peeking through gated walls at sleekly severe two-story houses that
were the latest style in 1967, their balconies overlooking yards that
looked only recently unmanicured. The neighborhood was alive
with workmen installing hot-water heaters trucked in from Pakistan, rewiring light fixtures and hooking up alarms. Aid organizations were taking over Wazir Akbar Khan, CARE and Worldvision
and Project Hope, cheek by jowl with the *New York Times*, the
BBC, ABC, NBC and *Newsweek*, renting out houses abandoned by
one of bin Laden's wives, by the Taliban police chief and by foreign
Taliban commanders.

But it was Afghan lawyers and doctors, engineers and judges,
who'd built those homes and furnished their kitchens with avocado-
colored appliances in more optimistic times. They'd sent their children overseas to study, preparing them for a world that still had not
taken hold. "We weren't perhaps as modern as girls in the United
States," Munavara had told me before I left Bishkek. "We didn't
date as freely and we didn't go to discos, although there were girls
who liked to go to Satara, the most popular nightclub in town. But
we certainly were comfortable talking to the boys in our classes,
wearing pants or short skirts. We were Muslims, but that didn't
mean that we were backward."

That Kabul had been obliterated, although it had not been erased quite as thoroughly as the Taliban had intended. Women displayed empty perfume bottles in their living room to keep its memory alive. Everyone was quick to pull out old photographs of parties and dances they had secreted in cubbyholes from "before." And that link to their past, and to the rest of the world, had been stoked by thousands of radios, television sets, VCRs and satellite dishes, hidden treasures that the Taliban never succeeded in confiscating.

I'd imagined that Kabulites lived cut off from the mainstream of the planet, and that they'd know little about international politics, hit movies and the latest rock stars. But virtually everyone—women and men, rich and poor—had kept an old radio at home and listened to music from Pakistan or news from the BBC. Those with some cash, like Jamila, watched VCRs—pornography the movies of choice among young men—and televisions hidden underneath platform beds, in abandoned outhouses or in closets. Satellite dishes were pounded into metal roofs, folded into trees, pulled out late at night in a city without streetlights.

The movie *Titanic* had been an underground sensation, a poignant irony considering how far Afghanistan sank, and even before the Taliban fled south, you could buy *Titanic* rice, *Titanic* shoes and of course, from behind the counter, hidden from official view, *Titanic* postcards and posters.

Enveloped in the first moments of liberation, the prospect of rejoining the world, Kabulites threw a love fest for the few Americans in the city.

"Is it true that she's American?" a woman asked my translator, poking her head out of her apartment doorway as I was climbing the five dank, unlit flights of steps to Nadia's apartment for the third time in as many days. When he told her I was from New York, she pulled me inside and kissed my hands.

"Thank you, thank you," she said, weeping. "Thank you for saving us."

I was taken aback. You don't expect gratitude when your tax dollars have just paid to bomb someone else's country. But the women all wanted to talk about the bombing, about the moment of liberation.

"We knew the Americans were going to bomb us, and of course, we were worried," Nadia explained. "We thought that it would be like with the Russians, with bombs falling everywhere."

Taliban leaders had scattered their antiaircraft missiles throughout residential neighborhoods, moved out of their homes and into the dormitories at Kabul University—maximizing the danger to the rest of the population. Thousands of ordinary citizens fled to the countryside. Nadia stocked up on food and prepared to move her bedding into the hallway to protect her family from shattering glass.

"Every night, we waited," she recalled, her hands dancing elegantly as she spoke. "When we finally heard the planes overhead, we lay down in the hall, all seven of us, and put cotton in the children's ears. When the bombs finally dropped, I smiled, really smiled, for the first time in years.

"Later, I heard on the radio that the first pilot was a woman. Here, women don't have the right to attend school and become literate. In America, they learn to bomb accurately. I knew at that moment that the Taliban would be finished and we'd have our freedom back. I knew I would have to thank that woman pilot for what she had done."

Still skeptical, I asked Najib, our translator, to show us the bomb damage.

"You want to know how accurate the bombing was? Let me take you to my neighborhood." He directed the driver to a row of houses and pointed to the single ruined building in the center. "That one, the ruined one, was where the Taliban defense forces put their antiaircraft guns," he said, pointing. "Look at the rest. It's a miracle. Not a single one was hit."

Najib was accustomed to his role as a guide for slightly offbeat tours. He'd taken Dennis to visit the mangy lion and fifty caged rabbits at the Kabul Zoo and had helped me calculate the plummeting prices of *borqas* at the market of Lycee Miryam. The only time he'd flagged was when I told him I wanted him to walk the field at the soccer stadium with me and bring the scenes of the stonings to life. After that, my request for a tour of bomb damage was a piece of cake.

We swung by the entrance to the InterContinental Hotel, where the remains of crumpled, bombed-out cars still blocked traffic. "The planes caught a group of Taliban just as they were leaving a meeting that first night," he explained. In the center of Wazir Akbar Khan, he halted the car on a street lined with elegant villas. Two had been decimated. "Those were the homes of the police chief and his guards."

Two weeks before our arrival in Kabul, Nadia had awoken to the news that the Taliban had departed. She gathered with other women in the hallway of her building to weep in relief, then strode into her apartment, pulled down the heavy drapes and opened the windows. It was the first time she'd felt the sun on her skin or the wind on her hair in five long years.

Jamila was too bleary from a night of bombing to absorb the news that the Taliban had fled, too terrified that it was a cruel hoax, that they were regrouping in the hills and would be back by nightfall. Two hours later, an engineer from Radio Kabul knocked at her door. That morning, he'd arrived to find the studio empty and driven directly to Jamila's.

"Do you dare?" he asked her.

Jamila didn't skip a beat. "Let me get my coat."

At noon, with the Taliban just over a mile outside Kabul, Jamila took her old seat in front of the familiar microphone, took off her *borqa* and waited for a signal to begin. "I presumed that if the Taliban returned, they would not only kill me, but kill my family," she said. But when the engineer nodded to her, she didn't miss a beat. "Dear citizens," she announced in free Afghanistan's first broadcast of the news, "I congratulate you on the collapse of the Taliban."

As word of the Taliban defeat spread across the planet, everyone expected the men to line up to get their beards shaved and the women to stream out of their homes and burn their *borqas*. Thousands of men did shave, and they were still celebrating when we arrived, flying kites, blaring their radios, playing chess, building homemade satellite dishes. With warlords of a dozen different factions flying into Bonn to hammer out the shape of the country's interim government, they gathered at restaurants and teahouses to debate the agreement, brimming with confidence: UN peacekeeping forces would enforce calm, foreign governments would send food and money, Afghanistan would be rebuilt.

But everything in Afghanistan was about gender, so while the men cut off their beards, the women clung to their *borqas*, testing both the public and their own emotional waters only gingerly. While Dennis was off rejoicing with the men in a world I could not enter, I listened to a less optimistic forecast for the future.

"Who are those people negotiating our future—*my* future?" Jamila railed as she watched the news from Bonn one evening. Her body was steeled with passionate determination; her eyes blazed with pain. "Except for the king, they're all war commanders, semi-literate people, the same people who destroyed Afghanistan in the first place. Why wasn't there a single educated professional from Kabul at that table? We need educated men there. Educated men respect women.

"And why are the women speaking for us all women who haven't lived in this country for years? Those women heard about our suffering. We lived it.

"Why aren't we there? Why aren't our voices being heard?

"If George Bush was struggling against terrorism in the United States, I was the one who struggled against terrorism in Afghanistan. I was a woman without a gun, without money, without an army. Mr. Bush had all of those tools. You judge whose struggle was more appreciable. You judge who deserves to have more say over the new government."

Even Jamila, who would blend in seamlessly with any group of tough New York professional women, wouldn't leave home without her *borqa*. "Women are afraid, and I include myself in that statement," she said. "The Northern Alliance says the *borqa* is now optional, but we have enough experience to know better than to trust them. We need a broad-based, multiethnic government. We need democracy. We need peacekeeping forces. Then maybe we'll be confident enough to take off our *borqas*."

Nadia, however, suspected that the root of the fear ran deeper than politics. "It wasn't easy to put on the *borqa*, and it won't be much easier to remove it," she commented, wistfully. "It's become part of our identities, and shifting identities is scary."

Nonetheless, subtle signs of change had turned Kabul into a pastiche of yesterdays and still tentative tomorrows. Although thousands of women clung to the known, to the safe, many others risked showing their hands by leaving their gloves at home. The old narrow pants worn under their *borqas* were giving way to wide bell-bottoms that flaunted a bit of ankle. Among the younger, bolder set, socks had even been replaced by patterned hose and closed flats by open-toed clunky heels that displayed brightly painted red toenails.

In the neighborhood of Lycee Miryam, the *borqa* shops—like all shops, metal containers left behind by the Russians—were half empty, the prices down by 50 percent. But at Safiullah Rasheedi's dress shop around the corner, business was booming, eight women—all covered in blue *borqas*—shopping for dresses for Eid al-Fitr, the festivities that mark the end of Ramadan. They held up mid-calf knit dresses with jackets, judging the size by sight in a world where trying on a dress was the equivalent of walking up Fifth Avenue naked.

Down Twelfth Street in Taimani, nineteen-year-old Laila Ahmadi stood outside deciding where to hang the sign being painted for her beauty salon. Under the Taliban, her shop had been illegal. Now, as she cut my hair—dry because she had no water—with big blunt scissors, a dozen women clustered around her, looking at her new photographs of movie stars and singers, planning the hairdos they'd wear for Eid. "Most won't do anything as drastic as they say they will," said Laila, laughing as she chopped off some more of my hair. "The fashion is to lighten the front of their hair just enough to feel modern. I don't think we'll see many bleached blonds under *borqas* this year."

She paused, then smiled. "But maybe next year."

Across town at Kabul University, women were trying on a different set of identities, but with even less optimism about their fit. The administration had announced that women would be permitted to study again when school reopened in the spring, and the first prospective female students had begun arriving to enroll. When I went over to meet them, I bumped into Maria, clutching a pair of black leather gloves in her hand, clearly ready to pull them on at the first sign of trouble. Her shoes were acceptably flat, but she wore lacy black stockings instead of socks. I don't know what she looked like because she was wearing a *borqa*, the first time I've interviewed a person whose face I couldn't see.

When the Taliban banned women from the university, Maria had been finishing her fourth year in the Faculty of Medicine, and for the next five years, she'd volunteered at an underground clinic that treated the women of a nation where they could not be touched by male physicians but where women physicians were not permitted to work. She'd taken a course in ultrasound in Pakistan and another one in television diagnosis. Mostly she'd stayed home and stared at the walls.

"I'm going to try to finish my studies," she said, her voice hushed. She'd only agreed to speak after leading me past a closed classroom door pockmarked with bullet holes and around the stairway to a discreet alcove. "But I'm not sure whether I'll be able to cope. I've tried to keep up. But my mind is not as good as it was. I'm not used to memorizing, and I've forgotten so much.

"I've been on Valium for four years and I can't concentrate very well. Besides, I'm getting old." She'd just turned twenty-five.

Of the 200 students who entered medical school with Maria, 82 were female. "Most have gotten married, and I doubt that their husbands will let them study. And few of the others have enough confidence left. I'll be surprised if even five join me."

She looked down, even though her face was already invisible behind the thick mesh of the facial opening of her *borqa*, and added, "I'll be more surprised if any of us make it."

Turning from the window looking out on the partially demolished university clock tower, Dennis reached toward Maria, then pulled back, remembering that touching a woman, especially in public, was a grave breach of propriety. "Oh, you'll make it," he said, attempting to encourage her. "You can do it."

Since—forgive the repetition—everything in Kabul was about gender, I was less sanguine.

During our two weeks there, Dennis and I might as well have inhabited two different universes. Muslim countries can be a delight for men raised in the West, where male bonding is a cliché about cheering on the home team together. In Afghanistan, men held hands on the street. They hugged and were playful with one another without a whit of self-consciousness, with the same easy affection women share in Europe or America.

Every time he stopped to change money—an enormous sack of bills for $10—to buy a can of milk or ask the price of a carpet, Dennis was instantly surrounded by fifty men of all ages and ethnicities offering him tea, inviting him home, braving their few words of English with warmth and hospitality. The guards at the Spinzar coaxed him into breaking the Ramadan fast with them, then handed him a rifle so that he could pose as one of them. His Kabul, then, was a city of open conviviality, of quick companionship and the offer of more meaningful friendship, a city that inspired optimism.

My Kabul wore a darker face. It was a city of silence and fear, an almost primeval world where women never quite knew what unseen force might hit them. The raw terror that still gripped Afghan women infected me, no doubt, but I would have felt the menace even without the contagion. It wasn't just that Kabul's public visage was almost entirely male, men outnumbering blue *borqas* by 500 to 1. Or that I couldn't invite my new friends to lunch or dinner. (Even the Herat, the most liberal place in the city, refused to serve Afghan women, despite my request.) Or that all traffic stopped the day that I drove a car through the streets, the open perplexity more antagonistic than amused.

It was the leering, the unabashed, uncontrolled lust of men catching their first glimpses of hair on the heads of foreign women. It was the sight of groups of men panting audibly at local women who dared expose their ankles or their hands. It was the unrelenting questions about sexual positions, about oral sex, anal intercourse, masturbation and threesomes with which Dennis was barraged by men already fantasizing about what they could do in a democracy.

Warped by five years of seeing female flesh only on porn tapes, the men were terrifying in what felt like unbridled lust. If they'd ever had it to begin with, they had lost their ability to take delight in the face of a beautiful woman or a slender waist. Looking at a woman in public was as illicit as watching pornography in private, and the reaction was identical.

It wasn't just the raw sexuality that fed my skepticism. I'd arrived in Kabul with the image of blue *borqas* scurrying down the street, unsure why scurrying seemed so fixed in my imagination. But scurry they did, and I understood why the first time Dennis and I took a walk. The streets were teeming with people, but while Dennis had no difficulty navigating the deluge, I was utterly unable to avoid being kicked, shoved, stepped on and knocked into. When I paused for a moment to watch the flow of pedestrian traffic, I got it, and was chilled by the implications: Women didn't exist. Men walked down the street as if the women simply weren't there, assuming, without even thinking about it, that the women would dash aside.

All the men I met hewed the politically correct line, agreeing that women's full rights should be restored with dispatch, but I was dubious, and more dubious still after a near-riot broke out after a group of nongovernmental organizations announced that they would hire only women as translators and aid workers.

Beyond gender, all those wonderful, warm and hospitable Afghans—and they really were, at least to us—seemed distorted in so many ways that it felt like a lifetime's work to untwist them. Our translator was a fourth-year medical student who had never cut open a body—and proclaimed that he could not under Muslim law. What hope, then, for the Afghan medical system? A law student told me that a jury system couldn't work because a Pashtun would never allow himself to be judged by an Uzbek, or an Uzbek by a Hazara—this, in the very same conversation in which he'd offered a definitive yes when I asked if Afghanistan could really remain a single country.

The women couldn't talk to the men, who couldn't stop looking at the women; the Pashtuns were out of sorts because they were being forced to share power with the Tajiks; the Hazara were reverting to banditry. And although everybody loved Americans in

November, history taught me that soon we'd be blamed for everyone's hatred.

We'd been in Kabul three days when I noticed a handwritten sign
by the entrance to the UN press briefing center: "CONVOY TO PAK-
ISTAN leaving Tuesday. The mayor of Jalalabad has promised us
troops and protection. If you're interested, leave a message for
Room 326."

That was my first indication that leaving Afghanistan might
prove to be more difficult than our "touched by a Russian angel"
entrance. The more I talked with other journalists, the starker the
choices became: The borders to Iran and Uzbekistan were still
closed, and the overland route to Tajikistan more dangerous than
ever. The UN allowed journalists on their flights to Pakistan, but
they wanted $4,000 a seat and had a two-week waiting list, that air
robbery justified by the high cost of insurance, a UN spokesman
contended. The only other alternative was to travel by land straight
east to Jalalabad, cross the Torkham border and make our way over
the fabled Khyber Pass to Pakistan.

Even when it was just a donkey track, the hundred-mile road to
Jalalabad was a hunting ground for bandits, ambushers, thieves and
warlords, and in the days of Kalashnikovs and four-wheel-drive
vehicles, it had become even more perilous. Just twenty miles out
of Kabul, past the hydroelectric dam that powered the city, the road
surface crumbled into a bed of stones and ruts and washboard dust.
On the left, the once-mighty Kabul River trickled through a narrow gorge; on the right rose the sheer Mahipar Mountains. Drivers
braked and veered in their efforts to avoid massive potholes and
one another as the road twisted and turned through the mountain
passes.

Before the Taliban imposed order on the chaos that was
Afghanistan, the most dangerous part of the road, near the mud village of Sarobi, was controlled by a warlord named Commander
Zardad who coveted the region not for its military value but for its
ideal conditions for ambush. Aid workers, food convoys and journalists moved regularly along that corridor, and Zardad's men were

there, either to extort protection money or rob them, depending on their willingness to cooperate.

The day after the Taliban fell from power, a bus was held up, the first robbery on the road in five years. Once journalists with computers, satellite phones, video cameras, still cameras and wads of cash became regular travelers, the road turned into a minefield of banditry. Just before we landed in Kabul, five cars of journalists had been stopped and stripped of their money and possessions. Several days later, three more were held up by thugs waving automatic rifles who threatened to kill them, allegedly with the permission of the local mullah. When the first convoy of humanitarian relief, forty-seven trucks sent from Pakistan by the World Food Programme, braved the road, the unarmed drivers were robbed of their personal possessions en route.

The Northern Alliance troops didn't even pretend to protect travelers from the bandits and from Taliban holed up in caves in the hills. The commander of the Pul Charkhi outpost, just east of Kabul, merely warned, "If you don't have a guard, they will steal everything from you."

The mayor of Jalalabad had taken to moving about with a heavily armed convoy, and the leader of one commando unit offered journalists the similar protection, for $5,000 per car.

Still, journalists made the five-hour journey, often with guards toting rifles and grenade launchers, because there was no other way in and out of Kabul. Then, just after 9 A.M. on November 19, reporters in a six-car convoy left Jalalabad for the capital. Twenty-five miles from their destination, two men flagged down the cars with Kalashnikovs. The driver of the first car hit the gas and managed to speed past, but the next two drivers halted. Six gunmen marched the four journalists up a hill, beat them, threw stones at them and finally killed them.

Frantic to avoid that minefield, we dropped by the office of the Russians who'd brought us to Kabul to see if we might catch a ride back to Dushanbe. Unfortunately, they were flying directly to Moscow, and the one visa we hadn't thought to get was Russian. They did have a helicopter coming in, though; at least, that's what an NBC fixer told us. "We've arranged to fly out with them, and you're welcome to ride along," he offered. "Be here tomorrow at 6 A.M., and don't tell anyone else."

Imbued with the optimism of the once-lucky, we pulled up to the NBC house before dawn. "Sorry, the Tajiks didn't let the Russians fly out again," the NBC folks told us. "That happened yesterday, too. We'll try again tomorrow."

Dozens of journalists were stuck in Kabul waiting for tomorrow—for the Russians to fly, for the UN to reduce its plane fare, for the military to put troops on the Jalalabad road. But with no working banks, no FedEx or DHL, no Western Union or Western embassies, we were all running out of cash in that all-cash society. After paying our driver, hotel and interpreter, we had $700 left. Neither of us was thrilled at the prospect of waiting for a tomorrow that might not come until the following week. Nor did we believe that a convoy provided much safety. It boldly announced: Look at us, we're journalists carrying enough stuff to keep you in rice and tea for the next decade.

"Is the Jalalabad road dangerous for Afghans?" I asked Najib as we drove away from the NBC house.

"Not really," he said. "The bandits know we don't have anything worth stealing."

Dennis was already dressed in Pashtun garb, a multipocketed vest worn over a long shirt and balloon pants with a three-yard-long blanket wound around his neck, topped by the ubiquitous Northern Alliance brown wool hat with a rolled brim. Only his high-topped green leather shoes were a giveaway. In a *borqa*, anyone can pass for an Afghan woman, and I happened to have a blue *borqa* in my backpack in the trunk.

"How much will the driver charge us for a ride to the border?" Dennis asked.

Fifteen minutes later, Najib and the driver had exchanged their Western clothes for more neutral outfits and picked up scarves to cover their clean-shaven faces. We drove out of Kabul in a beat-up old brown Toyota, four Afghans going to see relatives in Jalalabad.

I've read that the scenery along the road is spectacular, but I couldn't see it from behind the thick mesh of my *borqa*. I might not have noticed in any case. As we wound down into the gorge, past the most popular turns for ambushers, around the curve where the journalists had died, I was too taut to admire the landscape, too steeled to be curious about the American fighter jets streaking across the sky, leaving plumes of smoke rising from the hills to our south, too lonely inside my blue bag to absorb the extraordinary moment in history I'd just witnessed.

Dennis and I held hands underneath my *borqa*. We needed to hold each other, to trade fear and reassurance with a glance or a smile. The gentle pats and quick squeezes left us were cold comfort.

Our driver careened down the road, swerving wildly to avoid potholes and boulders, intent on reaching the safety of Jalalabad as quickly as possible. At one point, he swerved far to the left to get around the hulk of a rusting tank and sideswiped a taxi headed west. Our interpreter motioned us to silence as they climbed out of the car to work out a deal with the other driver, and to see if our car was still drivable.

Our luck held. Four hours later, we pulled up to the heavy metal gates at the border, a crush of trucks and cows and refugees trying to shove their way into Pakistan. I lifted off my *borqa* to the wide eyes of the waiting men, and without so much as a glance at our passports, we were ushered through by the immigration guards, my unveiled face all the evidence they needed that we were foreigners.

"Visas?" the Pakistan immigration police asked.

"We're journalists and need the emergency three-day visas your government grants journalists leaving Afghanistan," Dennis squeaked, the dust of the road having lent his voice an adolescent timbre.

The guards traded quizzical glances, then turned back to him. "What emergency visas? We've never heard of such a thing."

Dennis peered over his shoulder in dread, as if he could see back over the road to Kabul. The guards broke into spastic fits of laughter, then slapped entry stamps onto our passports and led us to a van bound for Islamabad.

We were too tired to feel euphoric or exhilarated at our escape, too fatigued from stress and, in a curious way, too saddened at our abrupt translocation from the extraordinary to the mundane.

That night, in a pristine hotel room in Islamabad, an eerily soulless immaculate conceit that allowed the Pakistani elite to imagine that they were somewhere other than a country on the brink, we took our first hot showers in two weeks and sent off a flurry of e-mails to reassure friends back home. Ice clinking in our glasses, unwrapped from their sanitized coverings, we turned on the television and paused at Fox News, broadcasting live footage of U.S. jets pounding a series of caves in Afghanistan's White Mountains.

Having been curiously cut off from news of what was happening in Afghanistan outside our view, we knew nothing about the final U.S. and Afghan assault on the caves we'd just driven past. But the scene on the television screen was instantly familiar from that afternoon's drive, the meaning of the jets we'd seen streaking across the sky suddenly clear.

The network cut away to the reporter. "This is Geraldo Rivera broadcasting from Jalalabad, Afghanistan, near the Tora Bora Caves."

8

Preoccupied Territory

The separated will be eaten by a bear,
the isolated will be eaten by a wolf.
—TATAR PROVERB

The U.S. Marines had landed by the time we returned to Bishkek, and the city was in a dither. Yanks, real Yanks, in uniforms and carrying guns, were setting up shop at an airport built by the Soviets. Before I could pick my way past the trash, around the wobbly piece of concrete that was my bottom step and through the three locks on my front door, I heard the news from the taxicab driver who'd driven us home from the airport, the woman who ran the kiosk downstairs where I bought my ninety-cent-per-pack Marlboro Lights, and two students I'd met in the courtyard outside.

"It's so exciting, they're everywhere!" one of them gushed. "They're in the bars and the clubs and all over!"

"Yankee imperialism!" the taxi driver had mumbled.

"Do you think they'll spend much money while they're here?" the kiosk owner had asked.

Facts were tough to come by in Kyrgyzstan since the local press didn't make much distinction between rumor and reliable information, leaving you with a classic Hobson's choice: you could believe the broadcasts of progovernment media, which printed the government line as the gospel, or you could opt for the version of reality transmitted by the antigovernment press, which assumed that everything the government said was a passel of lies. The truth, then, was a fuzzy concept, and people believed everything while claiming that they believed nothing.

Since I'd been wed to the concept of believing nothing since about 1965, the next day I sauntered over to the Pub and the Navigator to search for the rumored marines. A classic American hangout, the Pub was an old theater with twenty-foot ceilings, a dark wooden bar and waitresses with drop-dead figures. Offering the local interpretation of Mexican food—*mexicana sin* spices—they kept customers coming back by offering a pile of *Time* magazines devoted to the scandals of the Clinton presidency, a video library and regular performances by postmodern artists, the most recent, a Korean woman wrapped in a circle of spandex writhing to the grating sound of high-pitched drums. The scion of a Nepalese casino family who ran a local branch of the family business was drinking with a friend. And the sons of two bigwigs at Kumtor, the Canadian mining company that dug gold out of the hills above Lake Issyk-Kul, were tossing back beers at the bar. But they were toting golf clubs, not rifles, part of their conspiracy to bring civilization to Central Asia by constructing Bishkek's first eighteen holes.

With its glassed-in dining area, a retractable ceiling and removable walls that allowed for *alfresco* summer suppers, the Navigator attracted a mixed crowd of diplomats and gangsters, expatriates and students. A few stalwart missionaries and two embassy guys who I recognized from the town meeting were seated at wicker tables when I stopped by, but not a single marine.

I saw no foreign male with the requisite haircut at any of the clubs, at the bowling alley or at Tsum, the department store. I might have begun to suspect that the whole matter was yet another Kyrgyz media folly if I hadn't bumped into my travel agent, who informed me that she had trained some foreign soldiers in the use of satellite phones.

"Try the Hyatt," she suggested.

It had never occurred to me to look for U.S. military personnel at the most expensive hotel in Bishkek, but the minute I entered the bar, I knew that my tax dollars were paying for the gourmet fruit baskets that came standard with each room.

Who was the only journalistic question the guys I found, part of the advance guard, would answer. They were Air Force, not marines. For the *why, what, when,* and *how many more,* they referred me to the embassy. Alas, the public affairs officer wasn't yet "free to discuss the matter."

· Forced back onto the local media, I learned, depending on who I believed, or disbelieved, that 1,000, 1,500 or 5,000 troops were scheduled to arrive at a new base being built at the edge of the airport. They would be exclusively American, or a mixed European-American force. They would confine themselves to bolstering humanitarian relief efforts in Afghanistan, or they would drop bombs on innocent civilians in that war-torn land.

The only fact on which widespread agreement existed was that the United States would drop $7,000 into Kyrgyz coffers each time a plane took off or landed. The progovernment press touted that figure as an example of President Akaev's prowess as a tough negotiator. The opposition offered those same numbers either as proof that Akaev had been swayed into the American camp by filthy lucre, or as a reminder that the coffers most likely to be filled belonged to the president and his cronies.

Steeped in cynicism and passivity, Bishkekites were normally pretty indifferent to politics, a sport they found as convincing as, say, American pro wrestling. But the arrival of American troops— the first Western army in the region since Alexander the Great conquered Central Asia in 334 B.C.—proved irresistible. All of Bishkek had an opinion. Tanya, the owner of one of the "marriage agencies" that even my brightest and most attractive students considered their only way out of Kyrgyzstan, gushed that her business would finally take off, assuming American boys would consult her in order to arrange liaisons with local maidens rather than swing by the public bathhouse, where maidens of all shapes and sizes were readily available for 600 *som*, just $14.

Everyone else I spoke with was in a decidedly unfriendly lather about the new base, and I wasn't entirely unsympathetic, considering how I would feel if thousands of—say, French—troops landed at LaGuardia and began patrolling Seventy-ninth Street toting guns. I worried, too, that our boys might not all comport themselves like gentlemen and seek out Tanya's services, that the daughter of the wrong member of parliament would get knocked up, or, God forbid, raped.

As usual, none of my projections were echoed by my students.

"According to the agreement signed with the United States, the Kyrgyz militia will not have the right to inspect the planes that come in, so our country is about to be flooded with drugs," Rada declared with the surety of a person stating widely accepted fact.

Her logic escaped me. "How so?" I inquired.

"Everyone knows that America is filled with drugs," she responded with more of that same certainty. "We learned about it in school and we've seen all the American movies. All Americans do drugs, so of course the soldiers will bring drugs. And what's to prevent them from selling them?"

Should I parse out that logic? I wondered. Expound on the limitations of assuming that Hollywood movies reflect American society? Discuss random drug testing and military brigs? I opted for a different strategy.

"Why would they bring drugs with them when drugs are cheaper and just as available here?" I asked, scrawling "coals to Newcastle" on the blackboard to teach them a new idiom. "How much does heroin cost in Kyrgyzstan?"

The students feigned ignorance, or at least I thought they were feigning it, having no reason to suspect that any of them were shooting up in the filthy student bathrooms. But tons of heroin moved through Kyrgyzstan en route from Afghanistan to Russia and Europe, and it was common knowledge—so common that even I'd heard about it—that addicts spent about a dollar a day obliterating reality. Pot with an enticingly high THC content grew wild around Issyk-Kul and had become a major export crop for farmers unsure what to grow without direction from the Kremlin. You could knock on the door of any village house around the lake and buy a gram for less than 200 *som*, or $4. The year before, an elderly mother of ten was caught in Kazakhstan with ten pounds of pot she was taking to sell in Russia.

"Do you know how much these drugs cost in the United States? Junkies spend $100 a day on heroin, and good pot can cost $10 or $20 a gram," I continued. "So why would U.S. troops import them when they can buy them here?"

I was tired after a long trip home from Islamabad. All planes from Pakistan to Central Asia and China had been canceled for "the duration," as one travel agent put it, and he'd assured us that our only alternative was to fly home via London. We'd cobbled together a slightly more convoluted, albeit shorter route, catching a flight south to Dubai in the United Arab Emirates, a cab to neighboring Sharjah, and a second flight to Tashkent before taking a final plane into Bishkek.

Snow had fallen during our absence, and, at least to our Kabul-tainted eyes, the city seemed to gleam, just a few steps down from Las Vegas. That new romance, however, did not extend to my students. Despite all the warnings that I'd received about the passivity of young people in Kyrgyzstan, and all the promises I'd made with myself about lowered expectations, they were driving me crazy. Early on, I'd been awed by the breadth and depth of their knowledge, their ability to talk knowledgably about the Crusades and the Enlightenment, Shakespeare and Dostoyevsky. But untrained in critical thinking skills, they had no way to use their knowledge, to analyze problems, to unravel the loopholes in their own logic.

"Let me ask something else: If your militia were allowed to inspect the U.S. planes and found drugs, what would they do with them?" I asked them.

"They'd keep them and sell them," was the unanimous reply. No one disagreed. No one could. Dozens of new homes, virtual mansions, were being built on the outskirts of Bishkek and Osh, their saunas and three-car garages bought with the payoffs police and other government officials took from drug traffickers.

"Well, I don't like the idea of U.S. soldiers walking around downtown with guns," said Elvira, skirting the disconnect between Rada's concern and her government's corruption.

"Huh?"

"Well, according to *Vecherny Bishkek*, the agreement states that troops can be armed wherever they go, and I don't think they should be allowed to leave their area with guns."

"How do you know they will?" I had a hard time imagining an American general allowing his men to swagger down Soviet Street with loaded guns in their holsters.

"Why wouldn't they if they were allowed?" she responded.

Angela, who always had an opinion, had been atypically quiet. Suddenly she leaned forward and blurted out, "Americans are just too aggressive in their behavior. It makes us uncomfortable."

Since Angela had told me that I was the first American she'd ever met, I wasn't sure what to say. "Am I too aggressive for you?"

She stammered herself into silence, muttering, "Well, I don't mean you, but everybody knows that Americans are aggressive."

In the five remaining minutes before the bell rang, I was informed that the Americans wanted a base in Bishkek in order to contain China (imagine: one small base in Kyrgyzstan containing the Red Giant), that Bush's intention was to steal the old Soviet nuclear waste (steal it? Kyrgyzstan couldn't give it away), that Dick Cheney had his eye on Kyrgyzstan for an oil pipeline (the oddest of the theories, since Kyrgyzstan has no oil), and that little of the money that the United States spent, even on food and supplies, would help the Kyrgyz people, since it would all wind up in the pockets of the powerful.

They were undoubtedly correct on that last point. The contract for the airplane fuel for the base had already been awarded to Akaev's son-in-law.

My first-year students, at seventeen, the babies of the journalism program, voiced an entirely different set of concerns. "The Taliban and Al-Qaeda say they will attack any Muslim country that helps the United States," said a normally quiet girl, fear ringing through her broken English. "So this is a terrible mistake because it will make us a target."

There was no arguing with that theoretical possibility, but I wanted her to develop a more nuanced view. "Given the number of countries, Muslim and non-Muslim, helping the United States, do you think that Kyrgyzstan is likely to be their number-one target?"

"Why not?" a boy in the back asked, as if I had somehow impugned the honor of the republic.

"Well, let's think about it." I tried to be gentle, but I couldn't imagine a gentle way to tell them that their country wasn't important enough to attack.

"If you were the Taliban and wanted to punish a Muslim country for cooperating with the United States, which country would you be most likely to pick, which would you be angriest at, or which would be the best target: Pakistan, Uzbekistan or Kyrgyzstan?"

Blank stares.

"Why is Pakistan more important than Kyrgyzstan?" someone shouted out, pride addling his brain.

Gennadi, whose thick glasses would have made him the resident dork in America, raised his hand, then hesitantly read the statement he'd carefully penned in his book. Mortified when he made a mistake in English, in October he'd taken to scribbling down what he wanted to say before opening his mouth. "Maybe we have watched too many American movies, but we don't know how to act around professional soldiers."

"What professional soldiers?" I replied. "Our troops are all volunteers."

"Volunteers? How could such a thing be!" he exclaimed. "Why would anyone volunteer to be in the military?"

Kyrgyzstan's military was based on a universal draft, or at least in theory. In reality, you were exempt from service if you took an ROTC-like course at the university, or if you paid off the instructor to say that you had, or if you bribed the right person. And it was about to become even easier to evade eighteen months in a shack on the Tajik border because parliament was on the verge of passing a law that would effectively institutionalize bribery by waiving service for those willing or able to pay the government $450.

"But aren't there circumstances under which you would volunteer to fight for your country?" I asked my students.

The silence was profound.

"Okay, imagine that Uzbekistan invaded Kyrgyzstan in an effort to seize control over the Fergana Valley." I know, I'd used that scenario before, but it was apt in a dozen situations. "Wouldn't you volunteer to defend your country?"

"Maybe we're not so patriotic as Americans," Gennadi explained.

"Really, wouldn't anyone volunteer?"

Two girls raised their hands tentatively. The other twenty students scoffed.

I was standing outside, smoking another cigarette between classes, puzzling about why not a single student had so much as mentioned Afghanistan, when Vitaly asked if he could talk to me. The most earnest of my fifth-year students, graduate students, really, Vitaly always seemed to be struggling to understand concepts just beyond his experience, thus his grasp.

"Why did Russia agree to let the United States put a military base here?" His tone was plaintive as he wrestled with what he clearly felt to be an impossible, an inexplicable, reality. Although born in Kyrgyzstan, Vitaly was an ethnic Russian, and in the former Soviet Union, ethnicity inevitably trumped nationality.

"Why wouldn't they have agreed? And did Kyrgyzstan really need their agreement?"

Vitaly searched for a response I would understand. "But we're part of Russia. We're in the Russian sphere of influence."

"But you're not part of Russia, and you haven't been for ten years," I suggested.

"They are our big brother," he insisted.

"What has Big Brother done for you lately?" I probed the deep identification. "Do you think that abandoning this country to poverty shows filial responsibility?" After all, ten years earlier, Russia had dissolved the Soviet Union and taken its economic marbles home, leaving places like Kyrgyzstan bereft of even one working factory.

The need to identify with Russia wasn't a simple matter of ethnic identification, or even of a century-long habit. Being part of a mighty empire had made the Kyrgyz feel strong and important. Independence had reduced them to an insignificance they still couldn't quite swallow. Revisionism would have been too painful.

Vitaly was still trying to elucidate his angst over the base when his classmate Julia joined us. Already a member of Pyramid Television's reporting team, Julia wore her self-confidence gracefully. "I know others say they feel like the Americans are taking over, but the American troops make me feel safer," she ventured. Fond of skin-tight pants and skimpy blouses, Julia had taken careful note of the way Muslim fundamentalists treated women, undoubtedly the source of her hunger for protection. "We don't have such a good army," she added. That was quite an understatement. The Kyrgyz Air Force owned a single working plane. "Who else will save us if the IMU becomes powerful?" she asked, assuming both that Big Brother had taken a permanent leave of absence from familial responsibility and that someone else would step into that breach to keep Kyrgyzstan from sinking. "Even if the Russians were willing to help us, they're too weak.

"It's time for us to see if we're better off affiliating ourselves with someone else."

It took me several weeks before I wormed my way on to the new base, and it was neither the eagle on my blue passport nor my Fulbright professorship that opened the gates. As a favor to the embassy, I'd been helping a young TV reporter named Mamasadyk Bagyshov prepare for a two-week reporting trip to the United States, courtesy of the State Department. When he expressed an interest in filming at the base and interviewing U.S. servicemen stationed there for a piece on the volunteer army, I offered to set up the visit—and accompany him as a "consultant."

By then, local opinion about the base had pretty much polarized. Government supporters celebrated the windfall that Akaev's support for the global antiterrorism initiative had become—wasn't he clever at using the situation to the nation's advantage! (And it was some advantage—the Japanese had extended the government a $5 million low-interest loan for renovation of the airport, on top of the $46 million in credit they'd already given for repair of landing strips and cargo terminals. The Russians were providing advice on modernizing the national air-defense system, and the Chinese had extended technical assistance on upgrading the national security services. The United States was not only handing over an estimated $40 million for the base, but was on the verge of doubling its economic aid, a "reward" for the leasing of the base, most assumed.)

Dominated by Akaev's entrenched detractors, the opposition huffed and puffed with little coherence.

"Why do they keep talking about all the money we're getting when none of it goes to the people?" my students argued. "It all goes to the government. Anyway, the U.S. is only giving us that money because they want something from us."

"What's wrong with that?" I asked, forgetting that there's no Russian equivalent to the expression "win-win situation."

When a journalist from *Maya Stalitza*, an important opposition newspaper, met with staffers from the U.S. House of Representatives who came to Bishkek, she went one step further. "You're giving us too much money. You're seducing us. You're ruining us."

Older folks and younger communist die-hards unwilling to think of Americans as anything but slave drivers or capitalist lackeys whined incessantly: Look at them with their guns, they complained when local television stations ran film of soldiers patrolling the base perimeter, a four-mile circle that included several villages. Those Americans are so noisy and inconsiderate, with their planes taking off and landing all day. They're causing too much traffic, taking up tables in restaurants, creating lines at Tsum. They bought Akaev's soul with a pot of gold.

The disdain for Akaev's new alliance was seconded by Islamic fundamentalists, who churned out leaflets decrying the presence of the troops, stationed in Kyrgyzstan, they predicted, to impose godlessness on Kyrgyz peasants yearning to live under Shari'ah law.

Meanwhile, hundreds of young men waited outside the Hyatt—
which had finally reopened its bar, closed for lack of customers
after September 11—for a chance to apply for a decent job. And
mothers concerned that the future of Kyrgyzstan was bleak sensed
a moment of opportunity and encouraged Tasha or Sveta to go out
and find a nice American soldier.

Detractors and supporters alike lusted after a glimpse of the soldiers, for at least some minimal encounter to serve as fodder for
their arguments. Both sides were disappointed. The first order
issued by the commanding office, Brigadier General Christopher
Kelly, barred the troops from drinking alcohol in town or going
beyond the base or the front door of the Hyatt without a special
pass. Even that entitled them only to swing by the bowling alley
behind the Hyatt or the department store four blocks away. (Ironically, Kelly's alcohol order provoked more outrage among my students than any breach of their own freedom. "Men who work hard
have the right to vodka," they proclaimed.)

Located on the far side of town, out by the airport, the nameless
base was a thirty-seven-acre compound surrounded by a concrete
wall topped with coiled razor wire. Entrance was controlled by
armed airmen, and security was bolstered by four watchtowers and
guards with dogs patrolling in Humvees and on foot.

The namelessness was not accidental. The Kyrgyz government
had leased out that patch of the airport for a single year, and the
U.S. commanders were taking pains to reinforce the idea, or pre-
tense, of "temporary." But the troops, most members of the Air
Force's 376th Air Expeditionary Wing headquartered at Ramstein
Air Base in Germany, had already named their new home Ganci
Air Base, in honor of Chief Peter J. Ganci Jr., chief of the New York
City Fire Department, who died during rescue operations at the
World Trade Center on September 11.

General Kelly took pains to hammer home the "temporary"
message on television and in newspaper interviews, earnestly pointing out that his troops weren't even building permanent structures,
that the poor things were living in tents. When I think of tents, I
flash back to the Girl Scouts circa 1959, or a MASH unit during
the Korean War. But Ganci's tents—firm floors, vented heating sys-

tems that distributed the warmth evenly—bore as much relationship to Hawkeye's abode as a nylon Kelty with aluminum poles did to what I lugged on childhood camping trips.

The Americans at Ganci, who'd been joined by troops from France, Spain, Denmark, South Korea, Holland and Norway, weren't exactly roughing it. Within three months, they'd set up three recreation centers, a cinema tent, a post office, a sixty-bed hospital, a library, chapel, barbershop, laundry, cybercafé and fitness center. Oh, and the French, *bien sur*, had built their own dining facility, which offered fresh baguettes, a daily selection of cheese and French pastry.

The sight of U.S. troops working on a Central Asian air base originally built by the Soviets to protect themselves from U.S. troops felt slightly surreal, especially to those of us raised on Radio Free Europe commercials about Khrushchev threatening to bury our grandchildren. "I grew up in an age where this was indeed the big bear," said Kelly, former aide to Colin Powell. "And to have imagined even ten years ago that I would be in a former Soviet republic starting up an airbase and doing military operations was just inconceivable!"

The Kyrgyz officers echoed a similar theme, albeit with a slightly less upbeat tone: "But these are our enemies! What are they doing in our country?"

I wanted to believe, or perhaps it was a pipe dream, that the base was symbolic of a paradigm shift that most people were still too timid, or too emotionally invested in old truths, to acknowledge. Bush and Putin had just spent three days chowing down on catfish and pecan pie in Crawford and tooling around the canyons of West Texas in the American president's white pickup truck. And Vovotchka—the nickname of the old poster boy for the KGB—was planning to introduce George W. to the intricacies of caviar production (based, Putin instructed, on sturgeon cesarean sections) over dinner at his dacha.

The demise of the Soviet Union had pulled one leg out from under the old paradigm of the Russian Bear against the American Eagle, and it looked as if the war on terrorism might bring the whole outdated scheme of bipolarity tumbling down—unless hard-liners on all sides, white elephants from the Cold War and those with vested interests in keeping Great Game Redux alive, found a new twist to the plot. Old American warhorses seemed to be clinging to distrust of the "Rooskies" as a cozier, more familiar enemy in a world of unpredictable new ones. And suffering from post-superpower syndrome, the fiercely nationalistic Russian media were certainly fighting to maintain the competition, the only remnant of Russia's status as a world player.

"The authorities gave the best airport in the country to the Americans, as well as thirty-seven hectares of land for the military's needs," wrote *Pravda*, miffed at the inequity of the Americans' arriving just as the Russians were closing their last listening post in Cuba. "Nobody asked the common man if he wanted it or not, so the local population has to adapt to the new living conditions."

Many Kyrgyz parliamentarians, feeling trapped between the familiar, if less-than-generous ally, and the new well-endowed friend who used to be the old well-endowed enemy, echoed that dismay. "Kyrgyzstan, as a partner of Russia's for many years . . . is flirting with other countries against the interests of Russia," declaimed deputy Iskhak Masaliev. "When an ally makes some incomprehensible moves, it's an entirely understandable reaction.

"But now, what's done has been done. There's only one thing left to do: to get down on our knees and beg forgiveness from Russia, and to try and convince Russia that nothing like this will ever happen again."

9

Fear of Freedom

If a nation has a hero,
it will be safe.
—KAZAKH PROVERB

The heat came on a few weeks before I left for Afghanistan. It wasn't particularly cold out. In fact, the temperature had been rising all week, melting the five inches of snow that had fallen the weekend before. But it was November 1, and no matter how cold it was in October, or warm in November, by decree of Soviet state planners decades earlier, that was the day the heat was turned on all across the city, just as March 15 was the date it was shut off. Ten years into nationhood, Kyrgyzstan still hasn't reexamined the wisdom of centralized everything.

It wasn't just the fact of the heat. In good socialist fashion, it was the amount that was prescribed as well. Since the heat emanated from gigantic centralized boilers, every radiator in the city warmed up or cooled down simultaneously, and there was nothing anyone could do except open the windows or shiver under blankets because the radiator control knob had not been invented.

No one complained about the heat unless the Uzbeks shut off the flow on the natural gas line because the Kyrgyz government didn't pay its bills. The general unreliability of heat was considered to be an act of God, which was pretty much how the state was seen. The absence of gas, on the other hand, was an act of Uzbek greed, about which carping was permitted.

"Does it bother you that the government decides when you should be hot or cold?" I asked my students one afternoon when the sun broke through the winter chill, turning our classroom into a furnace.

"Of course it does," Dinara said, to the vigorous nodding of her group.

Three months earlier, that admission wouldn't have sounded like much of a concession to me, but I savored it as a triumph. Cut off from all contact with the non-Soviet world, the older generation never questioned the fact that the state had total control over the temperature of their homes any more than they questioned what had happened to all those people who'd mysteriously disappeared during Stalin's regime. Individual control of heat? Who ever heard of such a thing? Just another example of the way Americans let individual liberties run riot.

But the younger generation had watched too much Western television and seen too many Western movies. They, too, might not have been persuaded by the concept of individual rights. But they felt cold when the snow fell in the fall, and hot when the heat reflected off the snow of the Pamir Mountains.

They were not, however, angry. They had no inclination to anger.

"What can we do?" Dinara sighed in resignation, a sigh I remembered hearing from my grandmother when she was seventy-eight or seventy-nine.

"Write articles about why this happens," I replied. "Investigate whether the president and his cabinet have their own heaters. Look through the documents of the power company to see if they really are as financially strapped as they claim."

I could have gone on, but their eyes had already glazed over.

Week after week, they confronted the problems around them—the lack of heat, the grinding poverty, the corruption endemic to public life—with the same stultifying acceptance. I would have known how to cope if they'd been afraid of disappearing into some Stalinesque gulag, as some of their grandparents had, but they weren't. It simply never occurred to them to take matters, large or small, into their own hands. It never occurred to them to sport buttons reading QUESTION AUTHORITY. And it never occurred to them that there was anything strange about the fact that none of this had occurred to them. If rebellion is a natural teenage state, it had been successfully bred out of young people in Central Asia.

Public anger, after all, is a sort of responsibility, and in Kyrgyzstan people had been made too powerless by history, by what they had been taught as much as by what had actually happened, to feel comfortable carrying that burden.

Although the Berlin Wall had been torn down in an exuberant whirlwind of wreckage, the Russians had seceded from the Soviet Union, communism had been replaced with capitalism, and authoritarianism with what passed for democracy, still the people of Kyrgyzstan hungered for the reemergence of the status quo, as if those seismic historical shifts were just a temporary, momentary aberration, or at least for the appearance of a new status quo that wasn't so alien. All the new Western concepts floating around Central Asia—capitalism, individualism, democracy, freedom—felt too perilous and illicit to excite them.

One day I urged my friend Alfiya to introduce me to her mother, who I was tired of admiring from afar. A student at the American University, Alfiya, my backup translator, was a lanky Valley Girl with a Kazakh accent from a family—the only family I knew of—that was actually financially comfortable. When the Russians removed their subsidies and the economy went bust, both of her parents had lost their jobs. But her mother, a psychologist who'd always been interested in fashion, had reinvented herself. Once every month or two, she flew to Moscow and bought winter coats, which she sold in a stall at Dordoi, an enormous outdoor clothing mart that was a testament to the Chinese ability to mass-produce designer knockoffs. While most women of her generation were home with headaches, Alfiya's mother, having mastered capitalism, was clearing $1,000 a month.

"She won't meet you," Alfiya insisted. "She's too ashamed of what she's doing."

If trading was degrading, individualism was anathema.

"How did you do on your exam?" I asked a second-year student the afternoon after a calculus final.

"*We* did very well," she responded in GroupThink. "Two fives [As] and the rest fours."

Another day I tried to find out if Artyom liked Pushkin, a question that sounded ridiculous even to my newly tuned ears. No one in the Russian-speaking world disputed the brilliance or supremacy of Aleksandr Pushkin, the father of modern Russian literature.

"Pushkin is the most important Russian poet," he replied, reciting the phrase by rote.

"But you didn't answer my question," I continued. "Do you like Pushkin?"

Artyom fidgeted, caught between nineteen years of training and my persistent challenges, which had begun to prickle his skin. Recovering, he reverted to type. "If I don't like Pushkin, it's because I'm not yet educated enough to understand him."

That same politics of acquiescence pervaded Kyrgyzstan's nascent democracy, at least what the government called a democracy. A constitution had been adopted and elections were held regularly, although in the 2000 presidential contest ballot boxes had been stuffed, opponents harassed and popular challengers to the incumbent excluded by the courts. Kyrgyz "democracy" seemed safe only so long as the people continued to elect Akaev, as they had in 1989, 1991, 1995 and 2000.

I hadn't met a single person who liked Akaev, but they all admitted that they'd voted for him. Why? "The other guy might have been worse."

One day, when I was fed up with the constant grousing about the president, I urged my students, "If you don't like what Akaev is doing, impeach him. Circulate petitions. Demand that he change his policies."

I know, I know. I sounded like a broken record. But so did they with their typical response, "That's not our tradition."

"But this is a democracy, which means that the president works for you; he's your employee, just like George Bush is *my* employee."

By that point, they were relaxed enough to laugh in my face.

"In Kyrgyzstan, we work for the government, not the other way around."

American politicians would love Kyrgyzstan, I thought. No questions. No investigative media. No accountability. Everybody knew the government lied, but everybody accepted whatever it said.

"Maybe democracy is the right system for America," Regina was fond of saying, "but we're Asiatics."

In the West, at least in our uncynical moments, there's a poetry to freedom, a joyous hymn to human dignity, to fresh winds blowing out old indignities, to untethered birds and the "wretched refuse yearning to breathe free." It is etched on our minds as the images of Martin Luther King standing before 250,000 people on the Mall in Washington singing "We Shall Overcome"; Nelson Mandela, released after twenty-seven years in prison, declaring "Amandla"; thousands of Berliners tearing the wall down with sledgehammers, ice picks and their bare hands; even Billie Holiday wailing out "Ain't Nobody's Biz-ness If I Do."

The gospel of freedom isn't just political rhetoric; it's a quasi-religious belief that all human beings—blacks living under the boot of apartheid, women confined behind the veil in Afghanistan, Iraqis quaking before Saddam Hussein, Eastern Europeans and Central Asians girded for the knock at the door—suffocate without the right to think as they please, to say what they must, to roam where the spirit moves them.

Think as you please? My students in Kyrgyzstan puzzled over the concept. "What if you don't think the right thing?" Say what you want? "You can't have everyone running around saying just anything." Roam as you see fit? "But what about order? Safety and security? What if people want to do bad things?"

It's tough to wrap your mind around individual rights when you viscerally trust that only order, tradition, stability and the firm control of the State can hold back the chaos that is humanity. Imagine living your whole life in a house that is perpetually neat and tidy. The temperature is predictable, the furniture immovable, the curtains, like those on every house in the block, open to the street. One day you come home to find half the walls knocked down, some rooms frigid, the rest steamy, and nobody appears to tell you how to reestablish order.

Freedom, then, brought no euphoria to Kyrgyzstan. It provoked terror.

"This is terrible, that young people are having sex at such an early age and doing drugs," exclaimed Rada one morning. I'd assigned my class a reading about AIDS in the United States, planning to build a lesson around the structure of the piece. Rather, my students offered me a lesson in liberty.

"The State should not permit it," continued Rada, appalled at the statistics about thirteen-year-olds having sex and doing drugs. "They have too much freedom."

"What would you have the State do?" I asked. "Put seventeen-year-olds in the jail for having sex?"

"They need more religion," a Kyrgyz girl, a devout Muslim, theorized. "Religion will control them."

I blanched. "Should the State require families to be religious, and which religion should they mandate?"

"Well, the State is an extension of the family, so it has to act."

Tanya, my commie kid, interrupted. "I don't think the State can impose religion, but it has to do something to restrain freedom. If not, young people will die."

"What should it do?" I posed the question again.

"I think the problem is the media, so they need to control the media," said Angela, who then launched into a critique of reporting about the drug use of rock stars. "It glamorizes drug use. They shouldn't be allowed to do this."

"So should the State censor the media? Should the State prohibit the media from telling the truth?" That hit home. They were, after all, journalism students.

"I don't know, but I do know that this wouldn't have happened in Soviet times," Angela responded.

My favorite colleague at KRSU was a philosophy professor who refused to talk to me. "*Dobre ootra*, Clara Ajibekovna," I'd say, using her patronymic, when I saw her in the hallway.

"*Kak dila*, Clara," I'd chirp, inquiring about her health as we passed on the stairway into the building.

I suggested to the dean that we team-teach a class. Using students as intermediaries, I challenged her to debate. But no matter how hard I tried, I never succeeded in having a conversation with Clara Ajibekova, head of the Communist Party of Kyrgyzstan.

"She doesn't like Americans," my students explained when I told them how much I wanted to get to know Clara. "She hasn't ever talked to one and says she's not about to start now that America has destroyed the Soviet Union."

In the old days, Clara had served as the deputy head of the Ideology Department of the Central Committee of the Kyrgyz Communist Party and rector of its Institute of Political Studies. After the breakup of the Soviet Union, disenchanted with the reformist tendencies of the Soviet-era Party leadership, she had led the militants to form a new Communist Party, the true standard-bearer of Bolshevik revolutionary spirit. Clara favored Kyrgyzstan's joining the Russian Federation—not that the Russians had extended such an invitation—making Russian the state language, ending privatization and banning the sale and purchase of all agricultural land. In other words, she wanted to turn the clock back to 1990, if not 1952.

"Communism and the Soviet Union will rise again!" she declaimed regularly in her philosophy classes, swirling a cape or a scarf. A short, solid woman, Clara favored dramatic clothes in bold colors. "There are plans. Even as we speak, there are plans."

The students listened and laughed. Clara was the butt of scores of jokes on campus. But to many in the older generation, she was a heroine, the kind of strong leader whose clarity of vision could help Kyrgyzstan find its way back to the comforting familiarity of a strong State.

"Communism is the right system for Kyrgyzstan because it is based on respect," said Toktokan Edilbaeva, who we met during our search for a new apartment. After three months of being nickel-and-dimed by Vilena, our landlady, who had an annoying habit of forgetting to pay the electric bill, we'd decided to move. Toktokan owned a lovely two-bedroom apartment with hardwood floors, old-fashioned moldings and a perfect location on the edge of Pan-filov Park. Given its grandeur, I wasn't surprised to discover that she'd been a party official.

"We've studied the issue and realize what mistakes we made," she continued. I'd examined her library, which included all fifty volumes of the collected works of Lenin, so I had some notion of what she'd been studying. "We ignored what Marx wrote about the need to pass through capitalism before achieving socialism. So now we have to go through this period in order to rebuild."

The younger generation had less confidence in that illusion, although their vision of the Communist past was laced with a nos-talgic romanticism inherited from their parents and unmarred by so much as a dollop of revisionism. Even active prodding failed to provoke real skepticism.

At least once a week, Samarbek would mention in passing that "everyone was equal back then," until I finally couldn't resist chal-lenging his view of history.

"Didn't you say that your father owned three houses, one in your home village, another in Osh and a third in Bishkek?"

He agreed easily, sensing no contradiction.

"Did every family have three houses?" I asked.

"Of course not," he conceded.

"Then how can you say that everyone was equal back then?"

"But nobody knew about the houses in Osh and Bishkek," he answered flatly, as if resolving the matter handily.

Despite their unquestioning acceptance of every myth about "the good old days," young people maintained few illusions that Clara and Co. would conjure up a time warp to save them. But they were even more skeptical about the new form of deliverance being hawked by NGOs and foreign advisers: government "by the people." In their gospel, popular will didn't exist as a social, even as a revolutionary, force. Salvation could come only in the form of a great Messiah. All the masses could do was to wait and hope for his appearance, precisely as Christian fundamentalists wait for the return of Jesus.

I'd naively assumed that during Soviet times children had learned about the power of the proletariat, about the struggles of the masses, that they'd been primed, then, as had young people in Cuba for example, to do their duty as revolutionary activists. At least that's what I'd been taught in school.

But young Kyrgyz learned none of this; they'd learned about Lenin.

What brought about the revolution, which they still celebrated? Not the courageous anger of the masses. Not the historical moment, in Marxist terms. Not the brutality of the czar. It was Lenin.

"But Lenin could have done nothing without the will of the people," I argued, again and again.

"His genius was that he awoke and channeled that will," my students explained.

"But if he had been preaching that same message at another point in history, the masses might well not have followed him," I tried, hinting that mine was, in fact, the Marxist version of events.

"Oh, no," they wailed. "Lenin would have found a way to mobilize them. That's what a great leader does."

By the end of the first semester, the sight of a bevy of young women with a penchant for the Fashion Channel defending the genius of Lenin had become too discordant for my Western prejudices. I was about to leave for Iran on my winter break, so political idolatry was weighing heavily on my mind. I couldn't resist, then, taking on Lenin.

"What's with the Lenin worship?" I asked.

They were aghast. How could anyone need to ask why Vladimir Ilyich was worthy of adoration?

"But you don't know what he believed. You haven't read a word of his writings."

A chorus assaulted me: We don't need to know what he believed. We know what he did!

I admit that I still wasn't getting it. Maybe my brain had hit a pothole on the way to work. Maybe it just wasn't clear. But I was delivered by a quiet voice from the back: "Lenin, that's who we need in Kyrgyzstan now. Not Akaev, but Lenin."

Artyom, always the one who nudged me to understanding, added, "We're not like the Americans. We need a strong leader, someone with the strength to control us."

They weren't looking for an ideology, for communism or democracy, Islam or Christianity. With no money, no natural resources and a crumbling infrastructure, they weren't searching for a blueprint to build their future. Imbued with a national "can't do" attitude, they didn't have enough confidence to imagine being that powerful.

They were waiting for Lenin.

10

Mullahs on Motorcycles

The world becomes new every fifty years.
—KAZAKH PROVERB

It was the birthday of Imam Reza, the eighth Shiite imam, and the faithful poured into Mashad, Iran's holiest city, to pay tribute to the spiritual leader who'd succumbed to poisoned pomegranate juice twelve centuries ago. Each year, 12 million Muslims made that pilgrimage to kiss his gilt-edged tomb covered with a gold latticed cage, to pray at the adjacent Great Mosque of Gohar Shad and, in so doing, guarantee themselves places in paradise. The attack on the World Trade Center, the war in Afghanistan and mounting international concern that the mullahs in Tehran might be a tad disingenuous in pronouncing their disdain for terrorism might have decimated Iran's secular tourist industry. True believers, however, had not been deterred.

Neither had we, although we weren't making a bid for paradise, at least not in the Shiite sense. Before we'd left New York, we'd planned to spend our winter break following a portion of Marco

Polo's trek through Bukhara and Samarkand, in today's Uzbekistan, and east into Iran. After September 11, the prospect felt even more intriguing.

Had our goal been prayer, we could have joined one of the thousand religious tours that brought Muslims from across the globe. As infidels, however, we had to make our own travel arrangements, and they had been undermined from the start by the recalcitrant Turkmens—not to be confused with the Turks—who refused to give visas to foreigners crossing the Karakum desert en route to Iran.

Still living with one foot in the West, I thought, Okay, we'll just fly. But nothing's that simple in Central Asia, where all roads, trains and airplanes still lead to Moscow, some 1,500 miles north of Tehran. In the middle of one of my long nights on the Internet, though, I came across a vague reference to Aseman Iran Airlines, which supposedly flew directly from Bishkek to Mashad. Perfect, I thought. The next morning I stopped by my travel agent.

"Aseman Air? I never heard of it."

I don't give up easily, even faced with an airline that keeps itself secret. So when I couldn't find a phone number on the Internet or through information in Tehran, I turned to Samarbek for help. Equally intrigued and perplexed by the variety of complications two Americans could generate, Samarbek had become our Kyrgyz Guy Friday and he accepted the Aseman Airlines challenge with glee. For three days he made phone calls, dialing 109, local information, following leads and begging for help. We celebrated when he finally found a number for Aseman Charters. Alas, they offered service to Beijing, not Iran. But after a bit of wheedling, they provided him with a number for Aseman Airlines. On only the sixth call, a woman answered crisply. "Service to Mashad?" she said. "Of course, on Mondays, except when they change the flight to Tuesdays or Wednesdays. But we assure you there's a flight every week."

Iran, however, required Americans to travel with guides, and our agent in Tehran was equally insistent on a deposit. Nothing to it, right? Just fax him a credit card imprint. Except that American credit cards didn't work in Iran, a country that still hadn't apologized for snatching our embassy and its residents twenty-one years earlier. We dashed an e-mail off to our friend Billy in Rhode Island, asking him to stop by the bank and wire-transfer the funds directly into our travel agent's account.

"Are you trying to get me in trouble?" he shot back two days later. "It's illegal to wire money to Iran. I thought the bank manager was going to call the FBI on me."

One by one, we'd beaten back the bureaucrats, and on the eve of Imam Reza's birthday we boarded the plane, joining Muslims from all over Central Asia who jammed the aisles during prayer times to ensure a safe voyage. The next morning, we lined up at the security cubicles outside one of the four entrances to his shrine while the men were searched for weapons and the women were both patted down and inspected for appropriate garb. Elsewhere in the first Islamic Republic, an ankle-length coat and securely tied scarf satisfied the demands of the religious police. At the Place of Martyrdom, however, a full chador—an enormous black sheet that shrouds head and body—was compulsory for all women, and a bevy of female guards were on hand to oversee compliance.

Inside the vast complex—a wonderland of golden minarets, gilded cupolas and glittering mirrors—thousands of the faithful prostrated themselves on the cold concrete, their heads pointed toward Mecca. The stillness of the towering halls and serene court-yards was broken only by the nasal wailing of the muezzin, the mumbled chanting of the worshipers and the persistent clicking of hundreds of automated prostration counters.

Iran was a Modern Theocracy, so all the wonders of technology—and all the latest techniques of advertising—were employed to market and bolster Islam. No more old-fashioned stones against which you tapped your head during prayers. No more keeping mental count of how faithfully you'd fulfilled your religious duties. Keep track of how many times you have prostrated with the Mohr-e-Hazrat rak'at counter. No batteries necessary. Good in temperatures as high as 40 degrees centigrade—104 degrees Fahrenheit. Accurate up to 400,000 prostrations.

Forget glorifying the sacrifices of martyrs in old-fashioned verse or long-winded histories. Iranian martyrs were exalted on giant billboards with the same Technicolor overstatement used elsewhere to push cars or movies. The Koran was digitized in red letters that flashed on the type of huge electronic tickers that spell out the latest football scores in Times Square. Neon lights blinked out exhortations to prayer and to reverence for the Supreme Leader.

Even Imam Reza's shrine, one of the holiest places in the Shiite world, was so flooded with pink and green lights that it screamed Magic Kingdom to anyone who has ever seen Disneyland.

Elsewhere around the globe, fundamentalism had declared war on technology. In Iran, the two lived in apparent harmony, the result a sort of fundamentalist theme park, with the Ayatollah Khomeini cast in the role of a not-so-benevolent Walt. All the glitz and kitsch, the razzle-dazzle and seductively familiar iconography, made Iran look too modern, too American, to be taken seriously as a font of fundamentalist depravity. Evil, after all, lurks in hovels and caves among the illiterate and undereducated. At least that's the prevailing assumption. Iran bore little resemblance to the refugee camps of the West Bank or the Tora Bora Caves.

I'm not sure what I expected Iran to look like, just as I'm unsure what image I would have captured had we adhered to the standard letter M tour that the government expected visitors to follow: mausoleums, mosques, madrassahs, minarets, mullahs, muezzins and martyrs.

But we were more inclined to spend time in the market that sold used mullah cloaks than to examine yet another minaret. In Bam, we were happy to pass on the fourth ruined Zoroastrian temple if it would leave us time to drop by the office of the local religious police to see if they'd give us a copy of the poster of "acceptable female dress"—the four approved twists on the basic black sack with a hole cut out for the face—that hung in every hotel. And although the ruins in Persepolis, first capital of the Achaemenid Empire, were extraordinary, we preferred leaving an hour early so we could wander the remains of the postmodern Scheherazade complex—acres of air-conditioned tents with full plumbing and red damask reception areas lit by miles of torches—built for the famed $100 million bacchanal that precipitated the shah's downfall.

The Iran we explored didn't look or feel like any other place on the planet. It was a universe of jarring juxtapositions, dizzying ironies and impossible contradictions.

When Dennis first called the Iranian Embassy in Kyrgyzstan to inquire about a visa, he was put on hold and treated to Muzak renditions of "Dixie" and "When the Saints Go Marching In," courtesy of their American-made telephone system. And before I could even open the glossy Aseman Airlines magazine on our flight to Mashad, my fellow female passengers seated on the state-of-the-art Tupelov hastily concealed their blue jeans and bleached hair beneath long coats and black chadors, since even Iran's skies aren't friendly to the uncovered.

In Qom, still the heart of the revolution, mullahs in flowing robes and carefully wound turbans zoomed around on Japanese motorcycles even while debating the fine points of Koranic interpretation on cellphones. They flooded Internet cafés to post their latest sermons, even as they preached bans on public access to the World Wide Web.

The country was wallpapered with advertising, a growth industry once President Khatami lifted the lid on that old "capitalist evil." But since at least five percent of outdoor ads had to provide "cultural guidance," French perfume and Zim Zam cola competed for public attention with reminders like, "Rush quickly to your prayers," and the ruling couple—the president and the Supreme Leader—looked down from their lofty aeries atop major buildings onto billboards for Cacol, a candy adorned with the Coke logo.

Female villains dominated the silver screen in movie theaters from Yazd to Tabriz, but not a strand of hair peaked out from beneath their veils even when they waved their guns menacingly at the good guys. We turned on the nightly television news to catch the sports scores, and the newscaster sounded like any other jock recapping the game's highlights with breathless excitement. But in her triple chador (three layers of headcovering, in three different colors) she clearly had never been able to play any sport at all.

The sale and distribution of alcohol was banned, although Jews in Shiraz and Armenians along the Turkish border were free to brew wine and vodka at home—and did a brisk business supplying their Muslim neighbors. And the Ayatollah Khomeini, dead for thirteen years, had become his own industry, reproduced in dozens of factories on film and videotape, on calendars, date books, ashtrays, ceramic tiles, even Melmac plates.

It was enough to make me think the old man succeeded in creating a nation in which theocracy and modernism could coexist. And they did, to a remarkable extent, defying the power of the Internet (available, albeit slow); satellite television (illegal, but tolerated); international travel; and every word Marshall McLuhan wrote.

The mullahs who ran Iran had never shown much fear that the medium would undermine the message. In his war against Mohammad Reza Shah Pahlavi, Khomeini had wielded the audiotape as a revolutionary weapon, flooding Iran with dubs of his sermons. And shortly after he seized power, he reversed the decades-long fundamentalist hostility toward the movies as blasphemous iconography and threw his support behind the development of "Islamic cinema."

So in Iran, there was no contradiction between Stone Age fundamentalism and modern technology, between Gucci shoes and polygamy. There was no sense of irony whatsoever in the notion of turning the ayatollah's mausoleum into MuslimLand—think Graceland with domes, arches and minarets—or in dragging out Mickey and Bugs for a parade celebrating the defeat of the Great Satan and the establishment of the first modern theocracy.

That didn't mean there were no chinks in the system. On the street and in the bazaars, in teahouses, restaurants and hotels, from Mashad in the East to Tabriz hard on the Turkish border, the Iranians we met all complained, although rarely about the political stranglehold of the clerical elite, or the lack of political freedom. Rather, Iranians carped about the economy with the same blend of anxiety and disgust heard throughout the world, hardly a counterrevolutionary sentiment. To outsiders, a country with free education, government-funded retirement after twenty years of work, perfectly paved roads, superior health care and thriving commerce didn't exactly seem in bad shape. Iranians, however, had high expectations for their standard of living. Those expectations were not being met.

The leadership had evaded the rap by blaming the nation's economic problems on the Iran-Iraq War—which meant that they are entirely Saddam Hussein's fault, since Iran, of course, was the innocent victim of vicious aggression. (Saddam was the one issue on which Iranians were encouraged to agree with Americans, and people delighted at the thought that the U.S. government might use the war on terrorism as an excuse to get rid of their old enemy. It would be a double-barreled victory for Iran: the ouster of their greatest foe coupled with a delicious opportunity to condemn American imperialism.)

The eight-year bloodbath had devastated the country, forcing more than a million Iranians from their homes, polluting the countryside with land mines and sapping more than $240 billion from the nation's budget. But the conflagration had been over for more than a decade, and Iran's economic malaise owed as much to an unwieldy bureaucracy, state ownership of all major industries, crippling inflation and overregulation as to that carnage. Laying the responsibility at Saddam's feet, however, was a politically convenient ploy in a nation where no one would dare spit on the grave of a martyr.

"Do you think part of the problem might be the amount of money the government spends on religion?" I asked flatly. After two weeks in Iran, I'd concluded that I could ask anything if I did so without any inflection. I was desperate to gauge popular skepticism toward the religious establishment.

"Oh, that's not very much money" was the universal response. It was a curious blind spot. The government heated, lit and maintained hundreds of mosques, shrines and mausoleums, and kept building more, in Iran and abroad. All the mullahs in the country—and you couldn't go out without seeing the flowing robes of at least two mullahs per block—received government stipends. Then there were the religious police, who inspected everything from hotels to the piety of the army, and an enormous religious bureaucracy with veto power over secular decisions.

But Iranians were bombarded with a million-decibel blare of the message that Iran and fundamentalism were indivisible, reinforced by a thousand mausoleums, scores of *madrassahs*, a teeming mass of swaggering mullahs and millions of women shrouded in chadors clenched between their teeth as they dutifully followed the law and made their way to the back of the bus. Not even those who are old enough to remember what everyone euphemistically calls "before," then, thought twice—even once—about spending public money to support Islam.

The reformers, represented by the wildly popular President Mohammad Khatami, deftly finessed discontent by appealing to other facets of public vanity. After years of estrangement from the international mainstream, Iranians—proud of their long cultural tradition—chafed at their sense that Europeans and Americans might consider them to be backward. They longed to be hip, to be au courant. And Khatami indulged that yearning by fending off attempts by hardliners to shut down public Internet access and crack down on satellite dishes. Although he hadn't thrown out the old laws, he'd stopped enforcing regulations about alcohol, television, playing cards and videotapes that undermined Iranian illusions about their modernity.

Free to drink in their own homes, watch satellite television, even exchange porn tapes, then, Iranians basked in their coming of age and reveled in the thrill of breaking the law without fear that doing so would land them in jail.

And break it they did, with seeming abandon. In Kerman and Tabriz, we were repeatedly offered small cans of vodka on the street. In Shiraz, a group of teens approached us at an outdoor restaurant and unveiled a brick of hash on a table. Strangers bragged about the satellite dishes they had hidden in the garden or behind their laundry rooms, about their knowledge of the latest American films and their savvy about pop music.

One Shirazi hotel owner approached us as we were crossing the lobby. "Florida State, class of 1970," she said before bragging about how she smuggled books and magazines in from America. In the bazaar, women discreetly lifted their chadors to show off the latest fashions from Italy. Along the Caspian shore, we saw scores of over-size swimming pools on the grounds of the estates that lined the beach, although all were carefully walled in lest they attract the wrath of the religious police.

See, Iranians seemed to proclaim, we're sophisticated people. Iran is no backwater.

Yet:

One night in Shiraz, two local couples—the men civil engineers, their wives child psychologists—invited us home. We'd met over a special dinner and traditional music performance held in a teahouse carved out of an old caravanserai. Designed to provide the growing number of European tourists in Iran with tea, water pipes and a respite from the chaotic bazaar, the business had opened just two weeks after the World Trade Center attack. Until our arrival, they hadn't seen a single foreign customer.

"I can't let my children live in this country," one of the women told me between bites of saffron rice. Her voice was not hushed. "This country is a piece of shit." She and her husband had just received residency visas from Canada and were preparing for a summer departure.

"Freedom," she exclaimed, loudly, pointedly, when we arrived at her elegant suburban house and she slammed the front door, removed her scarf and shook out her long hair. Her husband pulled out a plastic gallon jug of vodka hidden inside a black garbage bag, clicked the remote that operated his satellite dish to one of the four Iranian stations beamed in illegally from California and invited our small gathering to dance—yet another publicly illicit activity. Their sons, older teenagers, chatted comfortably in English about their favorite video games, about the problem of drunk driving among Shirazi youth, and the popularity of hymen replacement—at $200 a pop—among middle-class women.

"Do you want to hear some Michael Jackson?" one of the boys offered. His father sneered. "Wasn't *Titanic* the most awesome movie?" he asked, clearly anxious to disassociate himself from his son's taste in music.

His companion poured yet another drink. "Don't Americans understand that bin Laden is being scapegoated?" he asked, adding a dollop of Fanta to six ounces of corrosive home brew. "September 11 was a Jewish plot. That's why no Jews were killed."

When we finally wobbled to the door, Dennis forgot himself for an instant and reached out to give our hostess a friendly hug. She

backed off in horror as her husband chided him, not so gently,
"Don't touch my wife. She's *my* property!"

Iranians of all ages, social classes and backgrounds greeted us
warmly and openly. "Oh, we're so glad to see Americans! We love
Americans!" They served us tea. They offered to take us home,
organize parties and teach us Farsi. When we stopped to photograph the DOWN WITH THE USA billboards at the entrance to the
town of Ardabil, between the Caspian and Tabriz, three youths
raced over to pose beneath it, their smiles welcoming, their fingers
split in the V of the universal peace sign.

But the warmth was laced with a curious historical blindness, a
fascinating blend of indifference, gullibility, hubris and self-pity.
Poor us, we're so misinterpreted and misunderstood.

How could anyone think Iran would be involved in interna-
tional terrorism? they asked. President Bush had not yet branded
Iran as a member of the troika of evil, but that indictment was
already in the wind, and Iranian newspapers were accusing the
United States of everything from warmongering to being a dupe of
Israel.

For Iranians, almost as galling as the indictment itself was the
prospect of being tarred with the same brush as Iraq. They might
have been divided over most political issues—from privatization to
the power of the clergy. But on the sinister nature of Saddam Hussein—provoker of wars, dispenser of poison gas, imprisoner of
innocent Iranian soldiers—Iran was united. Indeed, there was nothing George Bush could have done to provoke greater political con-
sensus among Iranians than to accuse them of behaving like Iraqis.

"Don't you think that Iran's encouragement and financing of
fundamentalist militants might have played some role in it?" I
responded. No one was willing to engage the question. "That's just
U.S. propaganda," they said, dismissively. "After all, we always
opposed the Taliban."

Journalists make lousy tourists. We're incapable of shutting up.

"What about Hamas and Hezbollah?" It slipped out before I
could bite my tongue.

"They're not terrorists, they're freedom fighters." The riposte
was pat. Automatic.

"Americans must be more respectful of Islam," ventured one teenager we'd met hanging out at Naghsh-e-Jahan Square—officially Imam Khomeini Square—in the center of Esfahan.

"Which Islam?" I pressed, almost sincerely. The Islam of Saudi Arabia, where women weren't permitted to drive, or the Islam of Iran, where they were? The Islam of Iran's Baluchi nomads, where women wore an elaborately embroidered rendition of a Lone Ranger mask, or of Kyrgyzstan, where anything went? Okay, so I was being impolitic. But it was getting tedious trying to keep the rules straight while being disparaged for respecting the wrong set.

No one wanted to argue politics. The shopkeepers, carpet salesmen, desk clerks and taxi drivers were too desperate to understand why we were the only tourists—literally, the only guests in 400-room hotels—in the country. "Why don't more Americans come to Iran?" they asked.

"Fear" was the only explanation I could offer.

"How could they possibly be afraid to travel here?" dozens of Iranians wondered, clearly wanting an explanation.

When I mentioned the fifty-two Americans taken hostage at the U.S. Embassy, one young man—like most Iranians, born after 1979—claimed that no such thing had happened. "You are clearly mistaken," he informed me, his voice grave and serious. Older folks, of course, couldn't deny that bit of history. But they dismissed the event offhandedly. "Oh, that old thing! That was a long time ago. Things here have changed."

During our last week in Iran, however, I made a pilgrimage to the old U.S. Embassy building—renamed the Den of Spies—at the corner of Taleqani and Shahid Mofatteh Streets in central Tehran. I'd read in the *New York Times* that it had been opened to the public with a permanent exhibition of the "Great Satan's Crimes against Humanity," complete with an Israeli flag doormat and a carnivalesque mock-up of Uncle Sam who visitors were invited to shoot with an air cannon.

Our guide, Hussein, a British-public-school wannabe, was not thrilled at my request to witness the other face of Iranian sentiment toward Americans, the "official" position of hostility toward the Great Imperialist. By then, I wasn't too pleased with Hussein either.

Our friend Meghan, a tall blond twenty-two-year-old American student, was with us, and we hadn't been in Iran twenty-four hours before Hussein, who was a full eight inches shorter, had declared his undying love for her. In a demented merry-go-round parody of *Romeo and Juliet*, Meghan, who clung to naiveté with near-religious zeal, alternately gave in to Ho's entreaties that she should stop by his room to "talk," and ran to ours when he expressed other desires, perplexed by his near-Shakespearean soliloquies. He would sulk, Meghan would feel guilty and make the ritual trip to his room, Ho's heart would soar, Meghan would tell him more stories about her boyfriend in Bishkek, and we were left wondering how many other foreign tourists he'd preyed on in his quest to find the perfect love route out of the country.

The embassy, then, became our first stop in Tehran, and as I spotted the compound from down the block, my mind flashed back more than two decades, filling in the now-peaceful street with images of angry mobs. I could still make out the outline of the eagle that once graced the lintel above the doorway, as well as the Great Seal hacked off a pillar at the entrance to the old redbrick structure.

The museum, however, was not open. In fact, I was informed that there is no museum. Perhaps, like the teahouse in Shiraz, it had fallen victim to the post–September 11 tourism bust. Perhaps cooler heads had prevailed. Even the Center for the Publication of the US Espionage Den's Documents, a shop peddling anti-imperialist souvenirs, had been closed.

But the walls around the compound still displayed Khomeini's warnings—*We will make America face a severe defeat* and *On the day the US praises us, we should mourn*—painted during those grim days. The colors weren't faded; the graffiti had clearly received a touch-up. In a new mural, added in 1998, Lady Liberty had been artfully transmogrified into a skeleton. She, too, showed no sign of wear.

part THREE

11

Pre-Prozac Nation

Better a titmouse in your hands than a crane in the sky.
—RUSSIAN PROVERB

Bishkek felt gray and lifeless after the neon-blinking color, the energy of traffic and the bustle of a country where people had both money and hope. The weather back home was damp, the clouds shrouding the mountains entirely. The heater in my new study wouldn't raise the temperature high enough to melt ice cream, so I'd taken to bundling myself in camel wool tights and the same green fuzzy cardigan every other aging babushka was swathed in. And I didn't think I could stomach one more roast chicken, our culinary refuge once we'd discovered that the Kazakhs weren't so literal about the concept of free range.

To make matters worse, I'd returned from Iran obsessed with hymen replacement surgery, a grim reminder of the plight of women trapped between tradition and modernization, and I could find no one willing to share my outrage. In most Muslim societies, a new bride and groom spent their wedding night with the groom's mother and sisters waiting impatiently in the next room to inspect the bloody sheet, and woe unto the bride who could not produce it. So thousands of urban, educated Iranian women were having their hymens sewn back up in the days right before their weddings, not to fool their new husbands, with whom they'd frequently had sex before marriage, but to appease their in-laws.

The war over westernization, modernization and globalization has long been fought over the length of women's skirts and the covering on their heads, in Christian, Jewish and Hindu societies as well as Islamic ones. What women do with their days and their bodies, whether they are permitted to drive cars, work outside the home, even *leave* home, has long been as much a political battle-ground as a private concern. In the shadow of the Taliban's reign of terror against Afghan women, hymen replacement surgery felt like a potent symbol of an enormous global struggle.

Outraged as only American feminists can do outrage, I vented my fury on Regina, my translator, for a full thirty minutes before I thought to ask, "This doesn't exist in Kyrgyzstan, does it?"

She examined me with that glance that unmistakably conveys *Are you out of your mind*? then said, "Of course it does. Surely American women do the same thing."

Almost speechless, I balked. "Surely they do not!"

Regina refused to believe me. "You just don't know about it!"

Rather than wallow in my funk, I concluded that I was suffering from second-semester-itis, that universal syndrome that afflicts college faculty all over the globe after the promise of the first term drowns in a sea of sloppy thinking, mediocre prose and drearily predictable students playing the same old drearily predictable games. At KRSU, things were no better, and on bad days I was convinced they were worse. Assignments were turned in late, if they were turned in at all, the 800-word stories I'd requested reduced to single paragraphs, or inflated, conflated as well, to ten-page exegeses on nothing.

Some days I showed up for class and found my room utterly empty. "Oh, sorry, we forgot to tell you, the students have a seminar they must attend," the registrar would inform me. On others, half the group would be absent, rehearsing for a Kyrgyz-language competition or a school play. Week after week, I was forced to defend my classroom from seminars, professional development workshops or wandering faculty members who had no permanent space of their own in the university's sparse quarters.

I was still butting heads with Katsev, my department chair, our last duel sparked when he'd asked if I might arrange to have some U.S. newspapers shipped over for our students to use.

"Given the cost and delay in shipping, surely it would be easier to arrange some extra Internet access for them," I proposed. "All U.S. newspapers are available online."

Katsev had scoffed, "No, no, they need to feel the paper and smell the ink."

I'd scoffed back, "This isn't Pushkin, Alexander Samuelovich. This is journalism. Lousy paper, shitty ink, and the next day you use it to wrap the garbage."

Even the key ladies had taken to scowling when I entered the building. I'd made off with my classroom key once too often, so they'd taken to accompanying me to my room and opening the locked door themselves.

To top it all off, I'd traveled 7,000 miles from a perfectly lovely existence in the Catskills. I'd called on every erg of creativity I could muster. I'd pushed, shoved, challenged and manipulated—and my students still showed less spunk than a herd of obstreperous yak lugging 1,000-pound yurts up the mountains.

By the end of the first term, I'd run out of patience with my third-year students, whose attendance habits had become as erratic as Kyrgyz mail delivery.

"You have to decide whether you really want to take this class," I finally told them. "It's your choice: Either commit your-self to coming and doing your work, or I'll cancel the class per-manently."

Eight pairs of eyes sparkled at the prospect of an extra free hour. They didn't hate my class, at least I didn't think they did, and they certainly weren't looking for more free time to suck down bong hits in the dorm. They were just permanently tired, a not surprising state given an absurd schedule that kept them in class twenty-five hours a week.

"Of course, if you choose the latter option, you will all flunk the course."

Before Regina had finished translating that message, the sparkles were doused by anger.

"That's not a choice," they groused.

"Sure it is," I snapped back, packing up my books and heading for the door. "It just might not be one that you like."

The following week, attendance remained spotty, and only half the group handed in their already overdue work. Could they really be that indifferent to their grades? I wondered.

Several days later, I was chatting with the registrar, Yanna, when the president of the third-year class entered, and with her, epiphany.

"Yanna, if I flunk all of my third-year students, will they offer you money to change the grades?" I asked, pointedly staring at the class president.

"Yes," Yanna responded without hesitation. We were friends, so she knew that I'd unearthed that not-so-dark secret months earlier.

"I'll tell you what, I'll double whatever they offer if you don't," I promised.

Miraculously, the following week, the entire third-year class appeared not just on time, but early. Their papers were stacked on my desk when I arrived, neatly presented, perfectly to length.

When I wasn't in a funk, I knew that I'd gotten through to a few. But the clock was ticking—I only had one semester left—and there were dozens more I needed to energize. I'd tried showing them the outlines of a different world, a new way of thinking. I'd taught them how to break down a story, conduct interviews and maintain balance. I'd challenged their worldviews, demanded that they stretch themselves—all to little avail. It was time for drastic measures.

If I'd been teaching in the United States, I might have resorted to all-out warfare, to threats, humiliation, or other forms of psychological torture, but my students in Kyrgyzstan were too fragile for shock therapy. Instead, I begged $500 from the U.S. Embassy to found a student newspaper.

By the standards of the United States, where every fourth-rate community college has a weekly rag, that ranked pretty low on the scale of revolutionary activity. But my students in Bishkek had never controlled anything of their own. They'd had no way to make their voices heard, no way to find those voices. I didn't expect our paper to bring down the government—God forbid, the embassy would have shipped me off to Guantanamo—but I hoped that the prospect of seeing their names in print might bring out a little feistiness.

At least, that was the fantasy that kept me going into the first week of the term.

It was tough to maintain that optimism when I was confronted with almost daily reminders of the failures of the other foreign advisers, teachers, engineers and consultants. And I was treated to the most vivid intimation of my potential defeat when our friend Meghan pulled up her skirt one afternoon and showed me an angry red welt the size of a pie pan with crimson tentacles shooting down her thigh.

A recent graduate of Carleton College, Meghan was a student Fulbrighter conducting research on women's groups in Kyrgyzstan. Savvy and fluent in Russian, at the age of twenty-two, she'd already traveled extensively in Russia, Mongolia, China and Vietnam. But like most young people, she took the bloom of good health for granted. So she'd ignored the festering insect bite on her thigh until she was well on her way to sepsis.

Samarbek, her boyfriend, had dismissed the wound as a boil caused by the cold. The son of a biologist and the brother of a physician, he nonetheless subscribed to the Kyrgyz belief that virtually all ailments are caused by the cold. Azamat had demurred and insisted that she'd eaten the wrong food, because everyone knew that boils were caused by bad meat.

Dennis and I demurred as well, suggesting that she needed to consult with a physician. In a country like Kyrgyzstan, you didn't easily utter the phrase "go to the doctor." The hospitals were filthy and cockroach-ridden, the equipment belonged in a museum of medical instrumentation, and patients lay untreated and hungry on fraying mattresses unless friends or relatives brought them sheets, food and medication.

But the U.S. Embassy had recently hired Kyrgyzstan's foremost cardiologist to spend her afternoons cleaning out the ears of the children of U.S. diplomats and dispensing prescriptions to their parents—an unnecessary conceit, since no prescriptions were needed in most of the pharmacies in town. It wasn't the type of medical work she was trained for, but the Americans offered her a weekly salary that was more than twice what most local physicians earned in a month.

The doctor whisked Meghan off to a "specialist" and we all breathed a sigh of relief. But two days later, the wound was redder, hotter and larger than ever. That's when we discovered that the "specialist" hadn't donned gloves before slicing a one-inch drainage hole into Meghan's thigh—although she had brought a pair for him to wear—and that he'd used a lance that had spent the day fermenting in a bowl of tepid water.

We rushed Meghan to a clinic run by two American doctors working in Bishkek as part of a program to upgrade the skills of Kyrgyz medical workers. Their diagnosis was immediate: Meghan's infection had gone deep into her flesh, the lancing insufficient to drain it and the initial infection compounded by a second bug, transmitted by the lack of sterility in the specialist's office.

"What's with the lack of attention to sterilization?" I asked a German friend. She laughed and shared the story of her encounter with a "specialist" who'd treated her son for an ear infection. "Don't you have any sterile materials in your office?" she asked when he started pulling cotton out of a box to create a Q-tip with his bare hands. The physician opened a cabinet filled with packaged disposable Q-tips and syringes, gauze, gloves and lances, all donated by the Americans, the Germans, the Swiss and the United Nations.

"They're entirely unnecessary," he said, closing the cabinet. "I know. I was trained in Moscow."

For a month, depression haunted me. It was etched onto the faces of the women who sat in the market for nine hours daily to sell six pounds of apples; marked by the heavy gait of the elderly women who knitted heavy wool socks sold for 60 *som*—$1.40—a pair to

buy food and pay their electric bills; woven into the frenzy of educated young women who signed up with overseas marriage agencies and Internet dating services, willing to sell themselves to
middle-aged strangers for a ticket out.

My reporter friend Mamasadyk, who'd provided me entrée into
the American base, had just come back from a month in the States
reenergized about life, and thus more disheartened than ever about
his future in Kyrgyzstan. At the age of nineteen, he was the most
intense Kyrgyz that I'd met—really, the only intense one—and that
personality quirk had been getting him into trouble since childhood. "I was the wrong type of person to live in the USSR," he
was fond of saying. He wasn't much more suited to his country's
new reality.

Born in a village so remote that you had to cross three international borders to get there from Bishkek, Mamasadyk had inherited
an independent streak from his father, a section leader on a collective farm who'd refused to stop praying to Allah, no matter what
the State decreed, and railed against the stupidity of the Supreme
Soviet's five-year agricultural plans. In school, Mamasadyk had tor-
mented his teachers with statements like, "I don't understand why
capitalism is evil." Then, while still in high school, he'd announced
that he planned to study in the United States.

His was a wild, crazy scheme in a community where no one
had ever been to the United States and where the collapse of the
Soviet Union and the breakup of the collectives had left everyone
impoverished. Newly independent farmers like his dad could find
no market for their produce. First his family lost their car, then
their animals, until they were left without enough money to buy
clothes for their six children.

How do you think you're going to get to America?
Mamasadyk's classmates had asked him, mocking his fantasy.

"They give scholarships," Mamasadyk explained boldly.

"Nothing in the world is free, especially if you don't have connections," his friends insisted. Local belief in conspiracies, hidden
agendas and inevitable corruption was near-universal, so seeming
generosity like U.S. government scholarships could strike no other
chord.

Mamasadyk had prevailed, but when he returned from a year in Los Angeles, he'd found Kyrgyzstan tough going. The American University had offered him a partial scholarship, but he couldn't raise the extra money he needed to eat and pay his bus fare. He'd won a full scholarship to Bishkek Humanities University, but quickly been advised to take a leave of absence after he raised a stink about professors accepting bribes for grades.

"In this country, if you don't play the game, you can't do anything," he said when he stopped by to talk about how to finish college in the States. "Unfortunately, I'm not very good at playing this game."

Of course, no one in Bishkek mentioned the word *depression*, as if everyone had silently agreed to pretend that rampant lethargy and hopelessness were normal states.

"Americans believe they have the right to be happy," our friend Gulnara laughed. "We have no illusions that we have any rights at all."

If the future looked glum for Mamasadyk, there seemed not a single ray of hope for Gulnara, whose permanent job was looking for work. Tall, slender and as graceful as Audrey Hepburn, Gulnara was a city Kyrgyz, the daughter of a staunch Communist, head of a collective farm on the outskirts of town. As a young woman, she'd followed in his footsteps and risen to a position of leadership in the Young Pioneers, the Communist youth group, and been honored with trips to meetings and conferences throughout the USSR.

But Gulnara was born with an aberrant gene that led her to question, an unpopular penchant in both Soviet and post-Soviet Kyrgyzstan. When she studied Materialistic Understanding of History 101—a required course at the university—she began to suspect that Soviet communism might not be what Marx and Engels had envisioned. When she turned in a thesis on the hippies, she was slapped down by her professors for using such a decadent word in her title. And when the lid flew off and communism collapsed, rather than mourn its passing with her comrades, she reveled openly in her sudden access to newspapers, magazines and books that had long been forbidden.

Her personal life was even more out of sync with the world

around her. Despite their militant communism, her family was 100 percent Kyrgyz, her father even having kidnapped her mother and taken her as his bride when she was a seventeen-year-old student at a teachers' college. Bride stealing was an ancient Kyrgyz tradition in which a young woman who'd captured the fancy of a not-so-young man—since most men set their sights on women a good twenty years younger—was swept up on horseback and kept overnight at the house of a prospective groom. In theory, she was free to attempt escape. But the deck was stacked against her. If she didn't succeed the first evening—escape being blocked by the groom and his drunken friends—it was pointless for her to resist. A woman who stayed the night at a stranger's home was considered "lost," tainted.

The custom had survived even modern communism. Like her mother, Gulnara's elder sister had been stolen, in what her husband and his friends thought of as a grand adventure. And one of Gulnara's classmates at the university was kidnapped by a young man she barely knew.

Gulnara, however, wasn't about to be stolen, or to mire herself in demeaning traditions. At the age of forty-one, she was single and had been unemployed for more than two years. But against all odds, she was intent on being a modern, progressive woman.

"People are holding on to tradition because they feel small, insignificant," she said. "They take refuge in the past. I just hope they don't try to take me with them."

Reeling from the onslaught of negativity that hit me when I returned from Iran, I sought relief in the classroom, excited to see the bright lights of my year, the smart, energetic fourth-year students.

"How was Tehran?" asked Elvira, the brightest of them all. Soft-spoken with a high-pitched voice that was almost inaudible, Elvira had lived in Tabriz for two years when her parents, both physicians, took work in Iran.

The night before, President Bush had made his famous Axis of Evil speech, which I thought lent my trip some added relevance in the classroom. Relevance, however, didn't interest them.

"Why do you like to travel so much?" Kuban, a Christian boy, changed the topic. A soft-spoken senior, Kuban had been drawn to religion by his need for an ideology that was "eternal, really eternal," unlike communism. He'd rejected Islam because it felt like the past. "Christianity is about the future," he said. "It's modern. Anyway, with Islam, there are all these rules about how you have to behave. With Christianity, faith is enough."

Daunted by the challenge of making sense out of why some people book cruises to Bermuda or spend the summer in Provence, while others opt for Bishkek—why some prefer the "right" places to the wrong ones—I'd responded to his question by launching into a discourse on the power of being a witness of history, one of the most compelling parts of my profession. Alexandra, a statuesque blond with long straight hair and a weighty air of entitlement worthy of Jerry Hall, interrupted.

"I hate hearing about your trips. None of us will ever be that free."

The room grew incredibly still. Tense.

"But we're just like American university students, aren't we?" asked Rada, revisiting an ongoing theme. Young people all over Central Asia craved reassurance that they weren't being left behind, that they were part of some barely perceived international mainstream that probably only existed on MTV.

"You're more depressed," I blurted out quickly, immediately regretting the haste of my mouth.

"I don't want to talk about that, it's too depressing," Rada replied, suddenly bursting into tears and running from the room.

The other students weren't sure who to comfort, but my horrified expression was staring them in the face. Rada wasn't.

"Don't worry, she's upset because she had a fight with her parents," Angela reassured me.

I should have known better than to crack the lid on that Kyrgyz Pandora's box, but my curiosity got the better of me.

"What happened?"

Rada's boyfriend was Kyrgyz, which made him off-limits to a Russian girl. Rada understood the rules of the game, but she'd still demanded an explanation from her mother, a physician at a Lake Issyk-Kul spa.

"He's Kyrgyz," her mother responded succinctly.

"And?" Rada had refused to give in.

"He's Kyrgyz."

The floodgates opened, and it all poured out. Angela, who was half Uzbek and half Russian, also had a Kyrgyz boyfriend. *His* parents had threatened to disown him—which meant not selling their horses and cows to help him and his bride get settled—if he married Angela. Elvira, who was Kyrgyz, was dating a well-educated and successful Georgian—from the country of Georgia, not the U.S. state. Her parents were barely speaking to her.

"Would you prefer me to get stolen by some illiterate from Naryn?" she'd asked them.

"Yes," they'd answered.

"It's not so bad for boys because if their wives aren't Kyrgyz, they always convert to Islam, so the parents can feel like they did something good by making more Muslims," Elvira explained. "Anyway, it's easier for a family to get rid of a wife they don't like than to get rid of a husband they hate."

These were my tough, confident girls, the female students with swagger and attitude. They were falling to pieces, right there in my classroom—and I couldn't muster a single word of comfort that wouldn't sound like I was playing Pollyanna. They weren't being melodramatic about being trapped in their personal lives. And the only jobs in town that paid a living wage were the handful with foreign embassies and the United Nations. They'd seen the future, at least on television and in the movies, but they couldn't imagine how they'd grasp it from Kyrgyzstan.

Finally, Alexandra shook out her long blond hair and sighed. "I'm going to Moscow to find an American. That's the only solution."

12

Olympic Illusions

*If you resent the louse,
you should burn the fur coat.*
—TATAR PROVERB

The banked lecture hall at the University of World Economy and Diplomacy in Tashkent, where both Hillary Clinton and Madeleine Albright had lectured students being trained to run Uzbekistan's foreign policy establishment, resembled Harvard more than Kyrgyz-Russo Slavic. Its staid formality only increased my apprehension as the dean steered me toward the podium. I'd dressed carefully that morning, in a suit and heels that had sat, unused, in my closet since I'd arrived in Central Asia. I needed all the ammunition I could muster to face what I'd been warned might be a fiery, contentious lot.

Uzbeks don't pride themselves on humility, the way Kyrgyz do. Notoriously feisty, they consider their nation to be a cradle of civilization, a leader of the nonaligned political world and an emerging economic power. Their best and brightest—all trilingual or better, all spotlessly well groomed, confidence oozing out of their pores—weren't about to take any guff from an American journalist, no matter that the dean had made her sound like a cross between Katherine Graham and Christiane Amanpour in his introduction.

Invited to address the students and faculty about the image of Central Asia, the image of Uzbekistan, in particular, in the U.S. media, I had decided to behave like an Uzbek, not to mince words.

"Before I begin my lecture, I want to pose three questions that I promise I'll get back to: First, what is the primary religion of Bolivia? Second, who is the president of Mauritania, or does it have a president? Third, what keeps the economy of Myanmar going?"

I skipped a beat, then continued: "As I was getting ready to come here and talk to you about the image of Central Asia in the U.S. media, I went online through the Internet into the archives, the electronic library, of the *Washington Post*, one of the most prestigious newspapers in my country, and typed in the name Kyrgyzstan. Between January 1999 and September 11 of last year, with the exception of one editorial condemning the unfairness of the most recent presidential elections there, Kyrgyzstan was mentioned only in passing, in news briefs and in a sentence or two in broader stories.

"The Kazakhs were better represented, although most of the reports dealt with the Olympics or the Russian space program. When Kazakhstan itself was the topic, readers learned about its president, who was called the 'Czar of Central Asia's largest country,' or about corruption, hardly a flattering portrait.

"I finished up my experiment with the word *Uzbekistan*, which received even more scant attention. Your nation was mentioned briefly a dozen times, usually in stories about the war in Afghanistan, the rising power of Muslim fundamentalists or international athletic competitions. The single piece dedicated solely to Uzbekistan was a report about the car bombing in Tashkent.

"And let me remind you again: the *Washington Post* is one of the best newspapers in the United States, with some of the most extensive and excellent foreign coverage of any American newspaper.

"So perhaps the most important thing I can say about the image of Central Asia in the U.S. media is that, until September 11, it didn't have one. Central Asia simply did not exist for most Americans, or if it did, it was just that vast space on the map to the west of China that used to belong to the Soviet Union.

"Before I talk about why this was, or what I think it means for Uzbekistan, I want to go back to the questions I started with. If you don't know what the religion of Bolivia is, and I suspect most of you don't, or who rules Mauritania, or what drives the economy of Myanmar, and if your media don't give you this information, why should our media do any better in informing Americans about Uzbekistan?"

The provocation was intentional, of course, so when the time came for questions, I gazed out at the audience, at the young men in carefully pressed jackets and ties, the women in tailored skirts and blouses, all of whom imagined themselves as future ministers and ambassadors, and girded myself for a smart, informed attack striking to the heart of U.S. foreign policy in Central Asia, if not in the Middle East and Europe. I'd boned up on how Madeleine Albright had dealt with President Karimov's overt persecution of religious Muslims. I'd read about the tangle of oil pipeline plans that, rumor had it, were America's true agenda in the region. I could detail, to the dollar, the score of U.S. aid programs that kept Uzbek educational institutions, Uzbek women's groups, Uzbek entrepreneurs, the Uzbek government, even the Uzbek military, alive.

I was convinced I was ready for anything when the first young man rose to a microphone, strutted actually, with the brashness of a young man who knew what he wanted to fight about well before I began speaking.

"Why did the United States cheat during the Winter Olympics?" he asked.

The Utah games had opened five weeks earlier to the usual flurry of optimistic rhetoric about global fraternity and the advancement of world peace through bobsledding, except in the former Soviet Union. The Russian nationalist press, still sore that a bunch of U.S. college students had humiliated the Russian ice hockey team in the 1980 Winter Games, whimpered audibly because that old U.S. team was chosen to light the torch at the opening ceremony, and I'd heard about that indignity every time I left my apartment.

When the French judge in the pairs' skating competition admitted to having made a deal to deprive the Canadian couple Jamie Salé and David Pelletier of the gold, my students, my neighbors and the woman who kept me in cigarettes had complained loudly, "Why do the American press lie about this?" And their ire flamed when Olga Koroleva lost to an Australian and two Canadians in the women's acrobatic freestyle skiing.

Then came the Larissa Lazutina affair, when the wildly popular skier was disqualified because her hemoglobin was found to be suspiciously high. "Don't they understand that she was menstruating?" earnest young men told me, as if well-schooled in the relationship between hemoglobin levels and menstruation. "The test was not done in the correct manner," others argued, without any details, since the Russian media hadn't told them what was incorrect about the manner in which the test was conducted.

Pravda opined that Lazutina was disqualified as revenge for her vociferous—some might say puerile—complaints about "fazed degenerates in uniforms who, in the state of antiterrorist psychosis, had crowded the locker rooms of Russia's female athletes, digging through personal effects," the Russian description of the routine screening to which all athletes had been subject.

"Just think of it, they found hemoglobin in Lazutina's blood!" *Pravda* continued. "Her blood was red! Of course, that could have been the result of the insufficient level of Coca-Cola in it."

When skater Sarah Hughes took the gold away from Russian favorite Irina Slutskaya, I was ready to lock myself in the house. Putin was denouncing the Olympic judges, members of the Russian Duma were urging their team to pack up and come home, pundits were demanding that the resignation of the Russian foreign minister—and all of that moral outrage was being heaped on the few Americans left on the streets of Bishkek.

After the Americans defeated the Russians in the ice hockey finals, I steeled myself for Putin to imitate his predecessor, Nikita Khrushchev, race to Utah and begin pounding his shoe on the stage next to the Mormon Tabernacle Choir.

Long after Kiss, in full metal regalia, closed out the games, the Russian media kept up the drumbeat. "Russia's opinion is almost ignored now, and the USA together with its NATO allies are openly imposing their will," wrote one columnist.

Pravda sounded an even more sinister alarm: "I dare say, deep in his or her heart, every Russian citizen has a bunch of scars on his or her national pride inflicted by so-called 'biased judges.'

"'Olympics as the school of hatred' sounds truly terrible. Yet this is exactly what the organizers of Olympics have managed to turn the games into by treating our athletes the way they do."

What was the agenda? *Pravda* asked rhetorically. "Gilding the idea of American leadership in everything without exception.

"May their objective be creating in 'these Russians' an inferiority complex, the feeling of being citizens of a 'redundant' country no one needs? May they achieve this forcing the Russians to endlessly and senselessly rummage through their resentments, in and out of sports, while the very taste of victory becomes forgotten?

"Is what is happening an attempt to substitute sports war for Cold War?"

The U.S. bombing of Afghanistan had provoked nary a whiff of anti-American sentiment in Central Asia. Even the opposition to the arrival of the U.S. Air Force had felt more like posturing than hostility to Uncle Sam. The Olympics, however, had ignited World War III.

The night before I'd left on my embassy-sponsored lecture tour of Uzbekistan and Turkmenistan, a bartender at my local Internet café had suddenly lit into America for its "arrogance."

"All this stuff about freedom and democracy doesn't mean you're so perfect," he grunted after an exegesis on the unfairness of the Olympic judges. The hostility had caught me off guard, since Sergei, a former member of the Soviet Navy, had always been warm and chatty.

"Is there anywhere else that's better?" Dennis inquired, as gently as possible.

"No," Sergei conceded without a moment's hesitation. "But Americans are too proud of their nationality."

Having been yelled at by taxicab drivers griping that George Bush had bribed the judges, railed at by students convinced the judges had been corrupted by pity for America, and accused, in some unstated fashion, of personally helping to reignite the Cold War, I'd been looking forward to spending the day at the University of World Economy and Diplomacy arguing about Afghanistan, oil pipelines and the Kyoto Treaty. But I couldn't leave the young Uzbek's question hanging.

"How do you know that the Americans cheated?" I turned back on him. "What's the evidence?"

The young man guffawed. "It's been in all the Russian newspapers, and I saw Larissa Lazutina on television stating that she had taken no drugs," he blurted out, as if he'd just pulled out a secret trump card.

"Do you think she'd tell the truth if she had cheated? And do you believe everything you read in the Russian newspapers?"

I didn't need to hear his response. By then I'd learned that while Uzbeks, like Kyrgyz, might be skeptical about what their own journalists wrote, they treated the work of Russian journalists with the instinctive respect obedient children accord to wise grandparents.

The young man stormed out of the lecture hall in full pique, ceding his place at the microphone to a faculty member. He, too, caught me off guard.

"Why does the United States have a plan to drop nuclear bombs on Russia, China and five other countries?" he asked.

I had no idea what the professor was talking about, and said so.

"According to your *Los Angeles Times*," he explained, "the Bush administration now has plans for nuclear attacks against Russia, China, Libya, Syria, Iran, Iraq and North Korea."

The light dawned. "Did you read the *Los Angeles Times* article?"

"No," he conceded. "But I read about it in the Russian press."

"Well, let me read it to you," I said, pulling the piece out of my briefcase. I don't normally carry around random *L.A. Times* articles, but I, too, had read reports about it in the Russian media—headlined THE USA IS GETTING READY FOR THE NUCLEAR WAR—and had downloaded the original for comparison. The "plans" in question were the "Nuclear Posture Review" conducted every six years by congressional mandate. According to the *L.A. Times*, the latest review set out plans for what the United States would do "in retaliation for attacks by nuclear, biological or chemical weapons."

"Why do you think the Russian press failed to mention that these were not offensive plans but retaliatory ones?" I asked after reading the whole article aloud slowly and with careful enunciation.

The young faculty member who'd brought up the issue shuffled momentarily, then smiled and replied, "They must not have understood the English correctly."

After a one-hour break for lunch, I returned to the podium to find a student waiting at the microphone.

"You asked why Americans should know more about Uzbekistan than about Bolivia. The answer is obvious. Uzbekistan is more important than Bolivia."

How did he know that? President Islam Karimov had laid it all out for him so frequently that it had become a local article of faith: Uzbekistan is rich in gold, oil, gas and endlessly fertile soil. For eight decades, Uzbek wealth had been pillaged by the Soviet Union to fuel the USSR's economic engine. With independence and, of course, Karimov's wise leadership, Uzbekistan could now take its place not only as the leader of Central Asia but as a leader of the world. So spake Karimov.

That cheery overview was the tip of the iceberg of what Uzbeks heard on television, on the radio and read in newspapers. Uzbekistan was, by presidential decree, the land of Good News, All the Time. The story line was clear and unwavering:

The economy is booming, the nation already almost self-sufficient in grain and gas. Investors like Daewoo, Mercedes-Benz and BAT tobacco are beating down the door of ministries in Tashkent for permission to invest. We're the world's second largest cotton producer, the seventh largest gold producer, and soon, we will all be rich. The President's Cup tournament has made Uzbekistan a Mecca for world-class tennis players, an Uzbek film won a prestigious prize at the Eurasian Teleforum in Moscow, and our president is not only beloved by his population—elected regularly with more than 90 percent of the vote—but is a world figure, the "Talleyrand of the East," whose counsel is sought by dozens of nations.

A few clouds darkened the horizon of the official portrait, of course. Egged on and funded by foreign powers, Muslim extremists were imperiling Uzbekistan's prosperous democratic future with their plans to impose a theocracy. But Karimov insisted that they would not prevail because he, the president, in his infinite wisdom, was rounding up anyone associated with those "enemies of the State" to ensure a peaceful and stable transition to democracy and capitalism.

"Uzbekistan is a country with a great future," declared Karimov. "Our sovereign, democratic state, which operates in accordance with constitutional law, is founded on the principles of humanism, and guarantees the rights and freedom of all its citizens, irrespective of nationality, religious creed, social class or political beliefs."

The unofficial story line went more like this:

An old Communist Party apparatchik appointed party chief by the Supreme Soviet in 1989, Karimov was elected president in 1991 and reelected in 1995 and 2000 with vote counts that proved him to be almost as popular as Saddam Hussein. Karimov played at democracy as if it were a grand game, creating, dissolving and banning a rotating series of political parties to create the illusion of an opposition. The charade was an elaborate dance that convinced absolutely no one since all four recognized political parties were kept so tightly under Karimov's thumb that the bills he submitted to the Oliy Majlis, the parliament, always passed unanimously.

When the Organization for Security and Cooperation in Europe (OSCE) criticized the last presidential election as corrupt, Karimov shot back, "The OSCE focuses only on the establishment of democracy, the protection of human rights and the freedom of the press. I am now questioning these values."

Uzbekistan's economy had shown real growth—Karimov wasn't making up that detail. But the country's oil and gas production was declining faster than new reserves were being discovered, and its agriculture was threatened by a high concentration of chemical pesticides and natural salt in the soil. Foreign debt was skyrocketing, and the International Monetary Fund had refused to restructure it because Karimov would not introduce a fully convertible foreign exchange. New foreign investment had fallen off precipitously when it became clear that currency-exchange licenses, which allowed overseas companies to exchange their *soum* profits for dollars, were revoked as soon as the profits became substantial.

The only organized opposition was mounted by Muslim fundamentalists bent on establishing a Taliban- or Iranian-style theocracy. But Karimov's predilection for blaming religious extremists for everything made it impossible to know just how many they were, or even how many had begun as nonpolitical Muslims who'd been pushed into rebellion by Karimov's penchant for rounding up willy-nilly men with long beards.

Highly skilled professionals were driving taxis, and villagers were struggling to eat. Corruption was so rampant that teachers openly solicited "tips" from their students, and many a deputy minister owed his august position to the bribes that he'd paid. But no one dared complain publicly since the president, notorious for throwing ashtrays and cellphones at subordinates who brought him bad news or dismissing underlings whenever he needed a scapegoat for another failed policy, brooked no opposition. Critics who hadn't already fled into exile were rounded up, beaten with planks studded with nails, subjected to electric shock and tried in courts that Stalin would have approved of, charged, of course, with being Islamic militants.

Yet in private, Uzbeks acted like Uzbeks, which meant they weren't shy about grousing. During a meeting I had with women leaders in Samarkand, they unleashed a torrent of complaints—about everything from the state of the economy and the pay of teachers to Karimov's decision to change Uzbek from the Cyrillic alphabet to the Latin. "That might not sound like a big thing to an outsider," said one elderly woman, her hair a shock of white. "But we've already been through the Arabic alphabet, the Latin alphabet, the Cyrillic alphabet, and now they are switching back to the Latin. Parents can no longer help their children with their homework, our libraries are becoming useless since everything is in Cyrillic, and, why? For the glory of the Uzbek nation as defined by President Karimov!"

At dinner that evening, in a private room of a restaurant plunged into darkness when the electricity failed, yet again, five Uzbeks regaled me with Georgian and Chuchik jokes, their brand of Polish jokes. By the fifth round of vodka, they were telling me stories about how many suitcases they had to fill with the virtually worthless Uzbek *soum* in order to purchase refrigerators or airplane tickets as the value of the local currency plummeted.

"This is the Uzbek model of a 'strong currency,'" one man quipped. "It's a currency that makes the people strong because they have to carry twenty pounds of it just to buy groceries."

Late that night, we drove over to Central Asia's most extraordinary women's shelter, a place so extraordinary that it was a national secret. The brainchild of an Uzbek physician, the shelter served the growing number of wives who broke under the pressure of traditional rural marriages—of husbands who beat them and mothers-in-law who enslaved them—and doused themselves with kerosene, then lit a match. Fire, they believed, against all Koranic teachings, was the only method of suicide that wouldn't keep them out of heaven.

The physician who cared for survivors whose faces were stretched taut by keloided scarring and whose shame confined them to the shelter should be an international heroine, I thought, honored by women's groups across the planet and supported by foreign governments, NGOs and feminist groups. But when I asked her to sit for an interview, she declined. "I can't let you write about me. Officially, I'm working with women who accidentally burn themselves on their kerosene stoves. If you print something that embarrasses Uzbekistan, I'll get in real trouble.

"People in Uzbekistan know that there are women burning themselves alive, but we're not allowed to discuss women being abused. The only bad news we're permitted to hear is about Islamic terrorism because that makes the president more powerful."

The Uzbek government had abolished prior censorship, leaving editors theoretically free to print other bad news. But they'd been warned that they would be held personally responsible for what they published, and at least three journalists were in prison—for reporting on spousal abuse, typhoid and government corruption.

Yet Karimov's critics are equally quick to defend much of what he did. The women's leaders in Samarkand heartily agreed with his crackdown on "fundamentalists," even though they acknowledged that hundreds of innocents had been caught up in his dragnets. And most Uzbeks applauded his decision not to plunge into economic reforms, convinced by Karimov that they would be consigned to the fates of their neighbors in Kyrgyzstan.

And they had succumbed totally, merrily, in fact, to the pride Karimov was instilling in the Uzbek people, a sort of Central Asian brand of manifest destiny. The founding father of that august future was Timur, called Tamerlane in the West, the Uzbek national hero. It was Timur who broke the Mongol hold over Central Asia, although he'd then gone on a nine-year rampage of looting and murder from Russia to northern India, from China into Iraq. Inside Uzbekistan, he was remembered for none of the bloodshed, of course. Timur, you learned at the Timur Museum and from every schoolchild on the street, brought a lawless society under control, promoted science, culture and the arts, and transformed Samarkand, his hometown, into a glittering capital.

Timur sat majestically atop his bronze warhorse in a square in the center of Tashkent where Karl Marx once held sway. Posters and paintings and statues of Timur were ubiquitous in public, and scores of private, spaces. Every bookstore had a major Timur section. And new operas and plays were being written to aggrandize his name.

Forget Russia, Karimov taught. We're Uzbekistan, the children of Timur. When the Russians were nothing but peasants led by petty princes, we were the people of Timur the Lame. We don't need the Russians, or the once-mighty Soviet Union, risen to power on Uzbekistan's wealth. We need no one, for Uzbeks will forge our own exalted future by following our own model of development, our own neutral path.

Karimov basked in Timur's reflected glory—what sense would it make to create it in the first place if you didn't plan a little politically convenient basking? And he laid out his program for rebuilding a country based on its great and ancient history in *Uzbekistan: Its Own Road of Renovation and Progress*, a national best seller, needless to say.

My last afternoon in Tashkent, I took a cab back to my hotel, past mile after mile of the same concrete-block apartment buildings that would tell a traveler who'd arrived blindfolded that he was in the former USSR. A few were being resurfaced, turned into cheap parodies of Madison Avenue, and a handful of buildings reflecting Uzbek architecture, white columns and turquoise domes, had gone up. Yet despite Uzbekistan's conceit about Tashkent—the largest city in Central Asia, the hub of Central Asia, the only Central Asian city with a subway—it looked as dreary and lifeless as Bishkek.

As we drove across town, my taxi driver tried to speak with me, although his English was only slightly more fluent than my Uzbek. After a mile or two of silence, during which he searched for the words, he asked, "You're American, no?"

When I confessed that I was, he went on: "What do Americans think about President Karimov's book?"

The lecture halls at the National University of Uzbekistan were not banked, and the students were neither meticulous in their dress nor sufficiently obedient to fit the mold at the University of World Economy and Diplomacy. The classrooms hadn't been painted in decades, and no one even noticed when the electricity blinked off.

I was scheduled to teach two seminars in the Faculty of Political and Social Sciences, a class led by a Dane working for the Civil Education Project, a Soros Foundation initiative. He'd prepped his students to be feisty and assertive, and the seminar turned into a free-for-all.

"I think the U.S. is scapegoating bin Laden," one boy almost shouted at me, as if I were the commander-in-chief, or at least the secretary of defense. "There's absolutely no proof that he was involved in what happened to New York."

"How do you know? Have you seen the proof?"

"If they had it, they would show it to the world," he responded. I was skeptical, but didn't press the issue.

"What about the videotape?" I continued.

"What videotape?"

"The one in which bin Laden essentially confesses."

"There is such a video?" he asked. What could he say? It hadn't been broadcast on local TV, and not seeing, in this instance, at least, was not believing.

"Tell us about Afghanistan," interrupted a Muslim girl, her hair discreetly covered, although Uzbek women were free to wear their hair open and loose. I launched into tales about the women I'd met, the damage they'd sustained, the grim task of rebuilding a broken nation. When I began to talk about how warmly we'd been received and mentioned the woman who'd thanked me for her liberation, a Russian boy in the class stopped me.

"Maybe she wanted something from you," he suggested.

The teacher smiled knowingly. The Soviet mentality—that nobody does anything without an ulterior motive—had infected even Karimov's younger generation.

I was surprised when the conversation veered toward Israel since during the six months I'd been in Central Asia, no one had shown the slightest interest in the situation in the Middle East.

"Why doesn't the United States do something?" an older boy inquired.

"What do you want the United States to do?" I really wanted to know.

"I don't know, but they must do something."

Another veiled female student broke into the conversation. "Who caused the problem in the Middle East in the first place?"

I didn't have a full hour to lay out the history of the founding of Israel, so I said, "Well, you could say the United Nations, since they created the State of Israel."

The woman shook her head vigorously. "No, no, the United States created the problem."

"How so? The United States didn't create Israel."

"The United States supports Israel," she replied.

Again, the conversation turned, making it impossible to have any real discussion. But talking with an American was a rare event.

"Why does the U.S. act only in its own interests?" one girl wanted to know.

"Doesn't the government of Uzbekistan act in the interests of the Uzbek people?" I responded. "Would you want them to put someone else's interests first?"

"Why doesn't the United States feed the poor?"

"Why doesn't the United States give Palestinians a state?"

"Why doesn't the United States stop the Chechen terrorists from murdering innocent people?"

"Why does the United States interfere in other countries?"

Hmm, I wondered. Don't interfere, but feed everyone and create world peace. Oh, without interference.

I took a peek at my watch: still thirty minutes to go. I was tired, tired of sloppy thinking, contradictions, Central Asia and my own dislocation. The television in our hotel broadcast the Star Channel, the first English-language entertainment we'd seen in months. I suddenly felt a pang of desire—a resistible urge, as it turned out—to watch Ally McBeal's brainless blatherings.

"Why doesn't the United States take care of the problems between India and Pakistan?"

I'd had enough. "Wait, wait, you want us to be everything to everybody, but there are limits to what we can do, even if we're the only superpower left."

The students froze, as if I'd just called Timur a mass murderer.

"No, you're not," one boy spoke for the group.

"We're not? What other superpower is out there?"

The group grew quiet until one girl shouted out, "The EU!"

Even the other students harrumphed at that notion.

"China's becoming one," another girl exclaimed.

"Becoming isn't the issue," I continued. "What other superpower exists today in the world?"

You could taste the reluctance, the unwillingness to concede that unique status to America. After six months in Central Asia, I'd begun envying the Swiss: all that money with no expectations or envy. But the Uzbek students were too far removed from any real sense of power—from confidence rather than bluster—to perceive its downside.

They searched for a retort, glancing nervously at one another and their professor for salvation. Then the boy who'd accused the United States of laying blame on an innocent bin Laden steeled his back and leaned forward triumphantly.

"Well, maybe not today," he proclaimed, "but in ten or twenty years, Uzbekistan will be a superpower."

13

Egomania-stan

It's the same donkey,
but with a new saddle.
—AFGHAN PROVERB

The Bosphorus Bar inside the Grand Turkmen Sheraton is 1,500 miles from the Bosphorus Straits. But in an age when you can eat fettuccine at the Adriatico Restaurant in Bishkek, sip a margarita at Los Amigos in the InterContinental Hotel in Tashkent, or boogie down at the Ipanema Club at the Hyatt in Almaty, Kazakhstan, few of the guests drinking Pilsner with their nachos on a Saturday night in Ashgabat, Turkmenistan, were giving much thought to geographical irony.

Rough-and-tumble Canadian oil drillers were too busy fending off a dozen local beauties dolled up in cowboy hats, boots and miniskirts, or jeans they'd grown into and never taken off, sending them over to swivel their hips at potbellied old codgers from Boeing, ExxonMobil, Halliburton or one of the score of multinational corporations looking to turn a profit in Turkmenistan, the westernmost Central Asian republic. A twenty-something Swedish-born Chinese consultant desultorily surfed the channels on the silenced television, pausing for the soccer scores from the BBC, moving on to the headlines from CNN and the Russian musical extravaganza du jour before settling on Al Jazeera, which was broadcasting interviews with relatives of the men imprisoned at Guantanamo's Camp X-ray between commercials for NileSat telephone service and the Bank of Oman.

Scheduled to deliver a series of lectures in Ashgabat, I'd just arrived from Tashkent, a convoluted peregrination that had involved three cars, a rickety old school bus, a half-mile hike and a ninety-minute, $2 airplane flight. Shortly before we'd left Bishkek, Uzbekistan Airways had canceled its weekly flight from Tashkent to Ashgabat in retaliation for the Turkmen government's insistence on turning over ticket receipts in *manat*, the virtually worthless Turkmen currency. By then, rearranging planes felt normal enough, and I'd booked a flight from Tashkent to Bukhara, a car for the two-hour drive from there to the border, another car to take Dennis and me from the border to the city of Turkmenabat on Turkmenistan's eastern border, and two seats on a plane from Turkmenabat to Ashgabat in the far west. It wasn't the best of plans for a woman who had always thought of herself as a direct-flighter. But by the standards of Central Asia, where a hundred-mile trip could take sixteen, bone-jarring hours, it didn't sound all that terrible.

Twelve hours before our departure from Tashkent, however, Uzbekistan Airways had cancelled their Bukhara flight. So just before dawn, we'd driven eight hours from Tashkent through Bukhara to a collection of rusting sheds manned by semicomatose guards in the middle of the desert: the border—sans towns, bathrooms or travelers. The Uzbeks stamped our passports in wonder. Why would any American go to Turkmenistan, of all godforsaken places?

We'd hoisted our packs and pounded down the unpaved road through the no-man's land to the Turkmen flag and a tiny bus, its paint faded, the cracked Naughahyde of its seats too tired to contain any stuffing. The fare for the 1.2-mile ride to the Turkmen border station was 25 cents. We were the only takers. The Turkmen guards read our passports as carefully as I'd pored through *Middle-march* on the night before finals. They didn't seem suspicious. Waiting for customers at a border crossing into a country loath to give out visas, they were bored, and more than a little curious. They even offered us tea while we languished in the middle of nowhere, waiting for the driver who'd take us to the airport.

The modernesque Sheraton was my first glimpse of globalization, Central Asian style, a tangle of tattooed truck drivers hauling food and medical supplies into Afghanistan, Malaysian businessmen, Swedish consultants, British engineers and a mob of Russian and Turkmen hookers. On the mezzanine two young Turkmen perched on stools strumming "Yesterday" on vintage Martin guitars, and then segued into a frenzied rendition of "Manitas de Plata" before ending their set with Bruce Springsteen's "Philadelphia," sung in Russian.

The kids who swarmed into Seattle and Davos to smash global capitalism would have been heartily reassured by the view from our balcony. Although the room could as easily have been in New Delhi or New York, outside there wasn't a single McDonald's, Kentucky Fried or Gap in sight. There was only Turkmenbashi, a homegrown product, the quirky, oddly anticharismatic Turkmen president, who looked like a used car salesman in an off-the-rack suit and a bad toupee.

Directly before us towered Ashgabat's signature monument: a fifteen-foot-tall golden statue of Turkmenbashi rotating slowly to follow the sun from atop a hundred-foot spire. It was colloquially called Three Legs, for the three pillars that formed its base, distinguishing it from the monument to Turkmenbashi called Eight Legs and the monument to Turkmenbashi called Five Legs, the latter a sculpture that also paid tribute to the great writers of Turkmenistan. (What, you haven't heard of them?) Only Forty Legs, a monument to Turkmen horses, did not explicitly honor the president. Never fear, an ample number of busts of Turkmenbashi nonetheless adorned the grounds around it.

No corner of Turkmenistan had been left free of bronze busts, marble statues, gilt monuments, brass medallions, billboards, photographs or oil paintings of the president. In front of banks and hotels, ministries, restaurants, shops and cafés, in the middle of the vast Karakum Desert and in classrooms across the nation, it was Turkmenbashi, his smile broad, his diamond and gold ring massive, all the time. At most intersections, there were so many newly retouched portraits of Turkmenbashi—a necessity once the president dyed his graying hair black—that the president was usually facing himself.

Turkmens could buy Turkmenbashi tea or cologne, watches, earrings or posters, all of which they paid for with *manat* emblazoned with his likeness. A Caspian Sea port city had been rechristened in his honor, as had a large meteorite that had smashed into the northern part of the country, the Ashgabat Airport, a textile factory, dozens of schools and the month of January, the month of Turkmenbashi's birthday, which was also National Flag Day.

Government officials wore small Turkmenbashi lapel pins, schoolchildren swore an oath to him every morning, and the Turkmen equivalent of the Pledge of Allegiance, a parody of the Jewish prayer, "If I forget thee, Oh Jerusalem," vowed the loss of a hand and a tongue for any betrayal of the president. The national motto was "Halk, Watan, Turkmenbashi," which translated as, "The People, the Motherland, Turkmenbashi." It was a cult of personality—narcissistic personality disorder writ large as political philosophy—that would have made Stalin blush.

The cult extended to the great leader's parents, long dead, his father in World War II and his mother in the devastating earthquake of 1948 that had leveled the city and killed more than 110,000 people. His father was commemorated with an enormous statue of a heroic soldier at the end of a long pink granite walkway. His mother had a park of her own, with a bronze statue of a woman cradling the young Turkmenbashi in her arms. The month of April bore her name, and the Turkmen parliament had awarded her the nation's top medal of honor, "Hero of Turkmenistan."

Turkmenbashi wasn't the president's real name. The former refrigeration engineer was born Saparmurat Niyazov and morphed from the head of the old Communist Party into the president before he anointed himself Turkmenbashi, the "leader of the Turkmen people." In 1990 he'd been elected with 98.3 percent of the vote, confirmed in office in 1992 with 99.5 percent, and prolonged yet again in 1994, by a stunning 99.9 percent of the electorate. He was the chief of the country's sole political party, the head of the army, the only officer with the rank of marshal, the president of the Association of Turkmens of the World and Leader of all Ethnic Turkmens and had been awarded the country's top medal of honor, Hero of Turkmenistan, five times.

Wasn't he embarrassed to see his face everywhere? A Russian reporter once asked him. "Why should I object if the people want it?" he responded.

All movement on the street stopped when our guide, Baxar, pulled her old Toyota away from the Sheraton into Ashgabat's meager traffic to begin our tour of what surely was the most surreal city on the planet. The eldest daughter of a traditional Turkmen family, in a loose, ankle-length burgundy dress with a patch of embroidery on the bodice and her hair covering her shoulders, she was the picture-perfect Turkmen woman. But after two years as a student in Kentucky and six months interning in Austria, Baxar, twenty-four, was willing to bend only so far to convention. Driving, an activity in which Turkmen women most decidedly did not engage, was the line she'd drawn in the sand.

As we sped down Gorogly Street, pedestrians glared in shock and horror. Dennis and I laughed, recalling the trouble I'd caused when I got behind the wheel of a car in Kabul: traffic stopped, and two drivers were so mystified by the spectacle that they'd slammed into one another.

As we wended our way out of the center, I saw armies of women with twig brooms sweeping the edges of major roads and shoveling out the gutters. Not a single male was engaged in manual labor.

"Where are all the men?" I asked Baxar.

"They consider that work dishonorable," she replied. So they spent their days in bars or teahouses, or scoring opium—130,000 addicts in a population of under 5 million.

The desert sun was blinding, although the air was tinged with a wintry chill as we joined the thousands of Turkmens who strolled around Four Legs or Eight Legs, bought *shashlik* from vendors lining miles of promenades, paused for ice cream by bandstands where Turkmen children twirled in traditional costumes and old men sang musical fables on long-necked two-string *dotars*. As city-proud as New Yorkers or San Franciscans, the residents of Ashgabat were dazzled by the monumental architecture, the pristine parks and elaborate fountains that Turkmenbashi had built to spruce up the old feature-less Soviet city. On Sundays they came out en masse to savor it.

Imagine a fourteen-year-old sitting home alone, playing a computer simulation game that allows him to design his own city. Then imagine that that fourteen-year-old boy is the president of Turkmenistan, with the cash to indulge fantasies about the marble facades of Dubai, the staidness of Washington, D.C., and the grandeur of the Great Wall of China. That's Ashgabat, a city of an ever-increasing number of hyper-lit buildings, fountains, wide boulevards, whimsical hotels, parks and trees—thousands of trees, all planted virtually full-grown, meticulously irrigated and maintained by work groups from factories, ministries and schools assigned, by the president, to keep them alive.

Downtown, the old public buildings that hadn't been razed had been resurfaced with white marble. On the edges of the city center, new apartment buildings reminiscent of early Michael Graves were springing up, their balconies overlooking new parks and fountains. And on the outskirts, a mini Las Vegas had been erected, a line of twenty-three hotels whose architectural styles were cheap flights of fancy, most with no guests and little plumbing.

In Ashgabat, appearance was everything, so the lack of plumbing and guests didn't seem to bother anyone, nor did the reality that the interiors of the marble-clad buildings were as decayed and depressing as their counterparts in Bishkek. No one seemed to care that the electricity didn't work, or that Turkmen women averse to standing over the squat toilets for fear that the draft might destroy their ability to reproduce tended to leave their waste on the edges of the porcelain. Or that the new apartment buildings stood empty because no Turkmen could afford $16,000 for a place to live. Or that the plethora of new fountains consumed so much water that most city residents were lucky to have anything flowing out of their taps for more than four hours a day. Or that thousands of city dwellers had lost their homes to Turkmenbashi's architectural pretensions.

"Isn't our city beautiful?" strangers asked. "Have you ever seen anything like it?"

We admired that beauty as we drove across town, everyone else's eyes still on Baxar. The final stop on our tour was the President's Health Walk, a twenty-mile set of concrete steps and pathways that snaked up the side of a mountain a few miles outside of Ashgabat, then followed the ridge line. It seemed to be Turkmenbashi's reinterpretation of China's Great Wall. No one had ever seen the president take a health walk. The day of the walk's dedication, foreign ambassadors and other dignitaries were expected to trek up from the bottom, a pretty steep two miles. Delivered by helicopter, Turkmenbashi had met them at the top.

But scores of buses and minivans disgorged Turkmens from towns and villages willing to brave the precipitous ascent. As I puffed my way up the incline, a gnarled, elderly woman—ninety-two years old—passed me, moving at a brisk pace, her son, in a traditional Turkmen hat that made him look like he was wearing an Afro wig made out of curly lamb, panting at her side. "Isn't it wonderful what the president has done for us?" she said chipperly.

As we drove back to the Sheraton late that afternoon, I tried to get Baxar to tell me what she thought of the rebuilding of her city. "Doesn't it bother you that everything is getting torn down, that people are evicted from their houses and left without water?"

She paused thoughtfully, then answered, "I'm not sure what to think. But every time I get angry, I remember: If I were redecorating my apartment, I'd have to destroy something.

"Maybe we have to sacrifice."

Back at the Sheraton that evening, we turned on the television, but even *Dallas* and Walt Disney had yielded to Turkmenbashi. On Channel 1 Turkmenbashi was exhorting his cabinet, whose members stood obediently, taking notes. On Channel 2, the leader graciously handed flowers to Turkmen dancers who were celebrating his greatness at a special telecast performance. And Channel 3 was running the Turkmenbashi miniseries *Turkmenbashi Is My Leader*, a nineteen-part film about the president's life and accomplishments, built around the story of an American woman journalist in Ashgabat who falls in love with . . . Turkmenbashi.

An hour later, Turkmenbashi was still starring on all three channels. The only break was the rerun of a corruption trial and "commercials" of happy Turkmen prancing across meadows.

Turkmenbashi's political style made Fidel Castro seem elusive and short-winded. No detail of public life escaped him, no matter how mundane, and he turned the most salacious of them into the nightly fare of the Turkmen viewing public in a Central Asian version of Reality TV. Hearing that his press secretary had been smoking in public—outside, that is, since smoking wasn't banned inside—the president had fined him one month's salary and demanded that he repent on air. Shortly before I arrived, he'd broadcast a public dressing down of the communications ministry for lack of good manners. "When I lift the receiver and say, 'Hello', there is no proper response or greetings to me," he complained.

At least once a month, viewers were treated to Turkmenbashi's demoting, berating or sacking a senior official who'd become a presidential threat. In June 2001 he'd ousted a deputy prime minister, in July the foreign minister, and in August he'd demoted the mayor of Ashgabat, the third most powerful official in the country, because there were potholes on the city's major thoroughfare. In January 2002 Turkmens watched as he dismissed two deputy prime ministers, the head of the Turkmen Railway Company, the chief of the fisheries commission, the minister of motor transportation, the minister of agriculture, the minister for the protection of nature, and the presidential adviser on legal issues, all for "grave shortcomings" in their work.

Even those spared public firing became stars in the daily drama of public humiliation. During my favorite episode—and it was hard not to think of what I was watching as a political soap opera—Turkmenbashi began with the head of the Tax Inspectorate. "You know that everyone pays taxes," he said, "but they [the tax inspectors] are collecting bribes rather than taxes. How many times have I talked about bribery in the customs service and demanded an improvement in the situation?"

He moved on to a provincial governor. "When you were coming here, I told you that I had criticized you on television. . . . Am I right?"

The governor bowed his head in contrition. "Esteemed leader, all that you said on television is right. I fully admit to all things you said, all my mistakes."

Next up was the governor of Lebap Region. "Now you, Berdimyrat, come here," instructed the president. "I criticized you for getting your daughter into an institute in Moscow and for giving her bodyguards and a car."

The Lebap governor demonstrated appropriate remorse. "All you said is correct. I have drawn the right conclusions as I see them. It will not happen again."

The president continued. "Where does your daughter study now?"

"She's now here," replied the governor. "She's not studying."

The president showed himself to be ever merciful. "Let her study by all means, but you must not provide such conditions— giving her two bodyguards and a Volkswagen to take her to her place of study. No Turkmen should do such things. Your child should study here instead of studying there. She does not need any guards."

Turkmenbashi wasn't done and turned next on the chairman of the Supreme Court, announcing, as if playing both Ed McMahon and Johnny Carson, "He proposed two candidates as deputy chairmen, and both of them are drug addicts. Do you think he doesn't know that? He knows that the people who are proposed by him are drug addicts."

Then he spoke directly to the chairman and his deputy. "You [the chairman of the Supreme Court] nominated three people. The third one, a Russian boy, is sitting next to you. I have learnt about all your shortcomings: you are a bigamist—why is your behavior so bad? Are you a deputy chairman of the Supreme Court or not?"

Turkmenbashi wasn't just a character in his own docudrama.

He was the people's choice, driving himself around in a mega-Mercedes, racing home for dinners delivered by an Italian-run hotel to the enormous gold-domed presidential palace where he lived alone. No one seemed to know what had happened to his wife and children, but the embassy told me that they had quietly moved to Moscow, allowing the president who built his rule around Turkmen pride to disguise the reality that he'd married a Russian Jew.

All was not entirely well in TurkmenbashiLand, although the bad news was not broadcast on television, where the president bragged that industrial output was soaring at world-shattering increases of 18 and 21 percent each year. Foreign analysts were predicting that the economy was on the verge of collapse, the treasury all but drained of foreign exchange. But even if they'd had any forum to broadcast their warnings, those analysts would have had a tough time proving their case since the budget figures were a state secret. Besides, the available numbers were calculated in *manats*, and it was never easy to figure out precisely how much a *manat* was worth because the Turkmen currency officially traded at 5,200 to a dollar but, unofficially, brought 20,000 per dollar. So if Turkmenistan's revenue was 20,562 billion *manats* in 1999, the last year for which anyone could find figures, was that $3.954 billion dollars, or less than $1 billion?

To make things even more difficult, almost everything in Turkmenistan was state-owned. So, how much did the Ministry of Health spend last year? That was hard to tell since they purchased their drugs from government-owned pharmaceutical companies and their electricity from a government-owned electrical supplier using government vouchers, and if they ran out of money, they borrowed it from a government-owned bank.

But the people loved their larger-than-life father figure. Unlike the Kyrgyz, the Turkmens didn't have to wait for Lenin. Turkmenbashi had already assumed his mantle, and he was putting on a great show. It didn't hurt that natural gas, water, electricity and the first fifteen minutes of local telephone calls were all free, that a gallon of premium gasoline cost less than a quarter or that a flight across the country was just $2.

Or that cable television had been banned after a Russian documentary showed starving Turkmen farmers, and that the leaders' opponents were not guaranteed equal time.

The director of the National Academy of Education scowled when I arrived to deliver a two-hour lecture on the U.S. education system to his senior research staff. I wasn't the problem. It was Maya, my translator, who made his lip curl.

"You can't translate into Russian," he told her. "You must translate from English into Turkmen."

Maya, the daughter of Russians who'd moved to Central Asia because it was "more bready," as she put it, spoke no Turkmen, so that instruction was nonsense, and the director knew it. "You translate," he instructed my bewildered embassy escort, a young Turkmen woman. He knew that her Turkmen was pretty rusty, and that it certainly didn't include words like *curriculum development* and *academic standards*. But he was not moved by her pleas.

The director wasn't being obstructionist. He was trying to avoid becoming a new star in the ongoing television firing drama. Turkmenbashi had decreed that all government-sponsored events were to be held in the Turkmen language, so my lecture turned into a linguistic farce as Maya translated my English into Russian and an academy staff member retranslated her Russian into Turkmen, although no one listened to that last part of the translation, since every member of the audience spoke perfect Russian.

Turkmenbashi had filled the ideological void left by the collapse of communism not with Islam, but with the glorification of traditional Turkmen culture. Old textbooks had been discarded and history rewritten to promote the pre-Russian "Golden Era" of Turkmen history. For those who needed tutoring in what that past meant for the present, Turkmenbashi had written the *Ruhnama*, a "spiritual constitution," as he called it—think a Turkmen replacement for *Das Kapital*. Encased in a shocking pink cover, the *Ruhnama* was a semipolitical stew: a summary of a thousand years of Turkmen history spiced with dashes of the Old Testament and the Koran, overlaid with a hefty dose of genealogy, presidential biography and poetry, with some newly coined Ben Franklin–esque proverbs and parables thrown in for good measure.

"The *Ruhnama* is the veil of the Turkmen people's face and soul," wrote the president in what was called a sacred text within forty-eight hours of its 2001 publication. "It is the Turkmen's first and basic reference book. It is the total of the Turkmen mind, customs and traditions, intentions, doings and ideals. . . . One part of *Ruhnama* is sky, the other part is earth.

"My basic aim in writing *Ruhnama* is to open the dwindling spring of national pride by clearing it of grass and stones and letting it flow again. . . . It is like replanting the arid lands of the past, which have become unproductive and useless, with the pine trees of the Turkmen plateaus. In this way I wish to rid us of the disease, trouble and anxiety of insensibility."

Need a sample? Here are Turkmenbashi's nine rules of good manners:

1. *Respect your elders.*

2. *Love your juniors.*

3. *Honor your father and mother.*

4. *Wear clean and decent clothes.*

5. *Keep goods at your home that have been earned by your own labor.*

6. *The decoration of the home, its order, cleanliness and appearance should be very good.*

7. *Protect the home and its exterior and neighboring areas and the place you live in.*

8. *Maintain sublime targets for your spirit.*

9. *Don't upset your wife or daughter and drape them with emeralds.*

Government employees were required to memorize passages from the national bible, government offices to devote an hour each week to its study, and students to tote Turkmenbashi's Big Pink

Book to their classes. With its translation into English, Turkmen State News Service was predicting "that the work of genius by the Turkmen leader would meet with wide welcome by the multimillion-strong English-speaking audience." The citizenry, stranded in a country with no foreign magazines, no satellite television, no Internet cafés and only a single, highly regulated Internet provider, believed.

While Kyrgyz have no real identity, and Uzbeks are being given a bloated one that looks like a recipe for disaster, Turkmens didn't have to struggle at all since Turkmenbashi was serving them up an identity on a platter: Happy Turkmen people living with glorious Turkmen traditions under the guidance of wise Turkmenbashi, according to his gospel, the *Ruhnama*.

If you didn't follow that path voluntarily, you were nudged along by the State. Students were obliged to wear traditional outfits to school and university. Government employees and their wives were subjected to background checks extending back three generations lest non-Turkmen blood pollute decision making. The national ballet had been closed because "there is no ballet in the blood of the Turkmens," according to the president. Pharmacists were instructed to "discourage" Turkmen women from buying birth control pills, which were dispensed freely to non-Turkmen ethnics. And, of course, everyone was expected to speak Turkmen.

That latter instruction was no mean feat since Russian had been the language of instruction for almost two generations and Turkmen had little vocabulary in medicine, science or much of any technical pursuit beyond cotton growing. Physicians suddenly required to chart in Turkmen were left scrambling for the words to note that a patient was suffering from a staph infection or needed a CAT scan.

Turkmen fever had been pretty hard on the non-Turkmens who are citizens of Turkmenistan. Russians like Maya were trapped by their citizenship in a country that no longer wanted them. Since they didn't speak Turkmen, they couldn't work for the government. They couldn't teach or practice medicine. But where could they go? Russia wasn't exactly begging for more unemployed professionals.

"It's not just the Russians who are suffering," Baxar admitted quietly my third day in Turkmenistan. She'd brought her boyfriend over to meet me, and her story flooded out over Cokes at the Bosphorus Bar, with Elton John as the audio backdrop. The gist of the problem: Baxar's boyfriend was Turkish. Since Turks are as close to Turkmen as you can get without being Turkmen, I had some trouble grasping the difficulty.

"My parents were hysterical when my brother fell in love with a girl from the wrong Turkmen tribe," she explained. "So imagine how they feel that my boyfriend isn't even Turkmen!"

Baxar's mother had encouraged her to study in the United States and to intern in Europe, but marrying a non-Turkmen was verboten—and she'd taken to bribery to bring Baxar in line. That ploy had worked with Baxar's brother, who'd succumbed to promises of a house and a new car if he left his girlfriend to marry his aunt. Tough and resilient, Baxar was not proving so easily dissuaded.

"What does your mother say? What's the matter with marrying a non-Turkmen?"

Baxar laughed. "Americans think that tradition is this romantic thing. But there's nothing romantic about it. What does my mother say? She yells, 'You are disgracing me! What will the neighbors think?'

"Tradition is just a bunch of gossip."

Over lunch that week, the dean at Turkmen State University leaned forward and told me, in the kind of hushed tone I associate with sharing a confidence about an illicit demonstration or a plot for escape from a brutal regime, "We would like very much to have more visits from American professors." He quickly surveyed the room and added, loudly, "It's not that we need help. We can do things ourselves. But it is interesting for our students."

Turkmenistan was the only corner of Central Asia, and one of the few corners of the world, where foreign experts were not particularly welcome. The history of Turkmenistan will be forged according to Turkmen tradition, Turkmenbashi had declared, which meant not only that Turkmenbashi and his minions spent their days reinventing the wheel, but that few outsiders were around to help the people see that the emperor was frequently nude.

That proud independence resonated with young people, stoking a nationalism that veered into arrogance. "We're Turkmen, we can do things ourselves," they said, as if oblivious to the reality that half the population lived below the poverty line, drug addiction was endemic and the countryside was dissolving in the dust of over-planted soil and saline water.

After lunch with the dean, I met with students from the history program at Turkmen State, a group of proudly opinionated young people entirely unlike my students in Kyrgyz. Our ninety-minute session on contemporary U.S. politics continued for almost three hours and turned into a brawl of anger, condescension and ignorance.

"What do you mean when you say that foreign policy issues are not at the heart of Americans' decisions about who should become president?" one young Russian shouted. "What is wrong with you people?"

A Turkmen senior took up the cry. "How can you say that your country isn't anti-Muslim when it only attacks Muslim countries?"

No one seemed to want to hear what I had to say, so I called on a demure girl in the back of the room, hoping for a respite from the contentiousness.

"You said that Americans are suspicious of very strong leaders like Turkmenbashi," she began. "Are they afraid of their greatness?"

Finally, a young man who had already attacked the United States for being anti-Muslim, going to war in Bosnia and sending troops to Somalia without seeing any of the contradictions in his position, asked, "Why do Americans always think every other country needs help?"

He paused, suggesting that he might really want a response.

"Many other countries request our help, our technical help, our financial help, our help with food and medicine and defense," I explained. The students' faces reflected uncertainty, perhaps even disbelief.

"Everyone doesn't need your help," interrupted the young man, working himself into a lather. "You're the ones who need help because you're delusional in thinking that everyone else needs you. You need psychiatric help."

14

Kent State the Kyrgyz Way

Cats don't catch mice to please God.
—AFGHAN PROVERB

I'd been a serious news junkie since Nikita Khrushchev traded shots of vodka and tequila with John Wayne in Hollywood when I was in the seventh grade, but I'd never heard the name Kyrgyzstan mentioned on an American television broadcast. Few journalists fight for the privilege of covering countries with so little cachet that stories about them are doomed to burial far inside the newspaper, somewhere between paid obituaries and the high school sports scores. With the foreign press corps in places like Central Asia minuscule, news from the region is, perforce, infrequent.

So when I caught the word *Kyrgyzstan* faintly intoned by a CNN anchor on the television across the bar at the Sheraton Grand Turkmen, I raced over and turned up the sound. It was just a one-minute spot, little more than a headline, but I extracted enough detail to know that I needed to phone home: Demonstrations in the southern district of Aksy had turned violent. At least five people were dead, sixty-two more seriously injured.

Trouble had been coming since January, when Azimbek Beknazarov, the fiery head of the Kyrgyz parliament's judiciary committee, was arrested and charged with official corruption. His alleged crime stemmed from his dismissal of a murder charge six years earlier, during his tenure as local prosecutor. His real offense was shooting off his mouth one time too many.

President Akaev had a long history of ridding himself of pesky opponents by throwing them in jail on trumped-up charges. He'd once ordered his former vice-president—who was also former minister of defense and former mayor of Bishkek—taken out of his hospital bed and hauled off to a cell because he'd allegedly arranged a shady sale of military hardware and abetted official forgery three years earlier. On another occasion, Akaev had sent a different leader of the opposition to jail for sixteen years for an alleged assassination plot despite the fact that the only witness against him admitted that the Ministry of National Security had ordered him to stage an entrapment.

The opening days of Beknazarov's trial had provided scant reassurance that another fix wasn't in. The presiding judge had summarily dismissed all of his lawyer's motions. Files from the original case had mysteriously turned up missing. And the alleged murderer had shown up in the witness box with a bruised and battered face.

In the weeks before our departure for Uzbekistan and Turkmenistan, the arrest and trial of Beknazarov had become a rallying point for Kyrgyzstan's weak and divided political opposition. A dozen or so supporters staged highly publicized hunger strikes. The opposition organized rallies in Bishkek. And parents in Aksy, Beknazarov's political base in the south, had begun keeping their kids out of school in protest. While the government hardly seemed on the verge of collapse, Akaev had managed the singular feat of turning a minor-league opposition figure into a national cause célèbre.

Beknazarov's trial, held in the Aksy province, had been suspended for almost a month when I left Bishkek for Tashkent and Ashgabat. When it resumed during my absence, 250 picketers showed up at the courthouse to demand his acquittal. The police and protestors battled for three days, the number of demonstrators rising in pace with the frustration and fury of the police. By the time the verdict was due, everyone with a complaint—from Beknazarov's arrest to the whimsy of the courts, the inequitable distribution of land from the old collective farms and the monopolization of power in the hands of powerful families from the north—was ready to march on the courthouse. A spark had been lit, I thought, and was about to explode as the people discovered their power.

March they did, in a phalanx 2,000 strong, until the police moved in to block them. Then a shouting match escalated into a scuffle, and policemen's clubs were answered with stones and rocks. By day's end, six protestors lay dead, victims of police bullets.

Before people around the country could even make sense out of what had happened, the interior minister was running for cover, alleging, alternately, that the police had been protecting themselves from an angry mob firing rifles and guns; that drunken hooligans had run amok; and that the political opposition had attempted a coup d'état.

Why hadn't his troops used rubber bullets or tear gas? "There's no money in our budget," he insisted.

The prime minister tried a different tack, denying that the demonstrators had been shot by police. The deaths were caused by knife wounds and flying bricks, he told reporters, without considering the possibility that someone might check the hospital records and discover that his version of events was a political fairy tale.

Then the president went on television and labeled it all an insidious plot by provocateurs and demagogues intent on subverting the legal system, part of a campaign by his opponents, who were too irresponsible to make mature use of their political freedom, to destabilize the country.

By the time we got home, thousands were massing in the south demanding the resignation of the president, and opposition political leaders were calling for the immediate suspension of the minister of the interior, the prosecutor-general, and the head of the national security service. The government had suspended the Beknazarov trial, sending the deputy home without bail. According to the news reports, bucolic Bishkek had turned tense.

I raced to class my first morning back, anxious to hear what my students made of what felt like the imminent collapse of the government. I'd imbued the first stirrings of dissent with a full measure of nostalgia about standing in front of a Woolworth's in Philadelphia in adolescent solidarity with students sitting in at Southern lunch counters, about the power of being one of a million voices chanting for peace on the mall in Washington, D.C., and about the miracle of Prague Spring. With the murders, I was transported back to Kent State and to Prague after the Russian invasion, overlaying Kyrgyzstan with my own remembered rage.

After eight months of listening to my students' passivity, I was dying to hear how they would sound with similar fury quaking through their voices.

"What's new?" I asked them breathlessly.

"Nothing," they replied flatly. "Everything is *normalna.*"

Normalna? I thought. Was everything I'd heard on CNN and confirmed with the embassy a fiction?

"What about Aksy?" I pressed.

"Oh, yeah, there was a riot and some people died," they responded with no trace of emotion.

"Some people say that the demonstrators were paid by the opposition to cause problems," one girl added.

"That's a government lie," countered another, without much passion or enthusiasm.

"How many people died?" I asked.

"No one is sure," they agreed.

"I heard an eleven-year-old shepherd boy was one of them," one boy interjected in the same tone he used to make excuses for late papers.

"How did they die?"

"The government shot them."

I heard no outrage in their voices, no special interest. Everything was *normalna.*

That evening, I brought up Aksy with Samarbek.

"This is good, what has happened, now maybe more people will demonstrate," he said.

"Would you demonstrate if you knew that your government

was willing to kill you?" I asked, fearing that the momentum for change had died with the protesters in Aksy.

"Oh, I hadn't thought about that."

The next day, I sought out Tanya, my student whose father was an opposition member of the parliament.

"This is the fault of the Americans," she insisted. "Because you want the base here, Akaev knows that you won't criticize him."

The argument was unsettling because Akaev hadn't received much criticism from the United States even before the building of the base. The darling of the Clinton administration, he'd been crowned the democratic prince of Central Asia. Hillary had come for a visit, and Madeleine dropped by. They couldn't heap money fast enough on the only president in the region who understood that doing a convincing imitation of a democratic leader was the key to U.S. largesse. In truth, it hadn't been all that difficult for Akaev to play that role since his opposition was bumbling, dominated largely by a few old comrades demanding the reunification of the Soviet Union and the restoration of communism. Free speech? Sure, since no one was saying much that Akaev didn't like. Free press? No problem, given that the government owned the only large printing press in the nation.

And Akaev certainly didn't have any competition for most democratic president in Central Asia. Nursultan Nazarbaev had turned Kazakhstan into a family fiefdom, installing his eldest daughter at the helm of the state-owned television network and her husband in the top spot of the tax office. His youngest daughter's husband ran KazTransOil and, as owner of the largest construction company in the country, was awarded virtually every government contract. The motto of Islam Karimov was "a strong rule, a strong state," which pretty much summed up democracy in Uzbekistan. And Turkmenbashi . . . there's no need to say more about Turkmenbashi.

But Akaev's record had hardly been that of a stellar democrat. During his decade in office, he'd shown a penchant for ducking sticky situations by firing officials, forming new governments and rewriting the constitution. When allegations of corruption against his closest political associates threatened to blossom into a major scandal, Akaev simply dismissed the government and called upon the last Communist premier to organize a new one, as if to say, "See, it's not me. I'm trying, but everyone else is a mess."

One year, when so many members of parliament boycotted the final legislative session that a quorum could not be reached, Akaev accused the Communists of trying to undermine the new democracy, although many suspected that the president had engineered the boycott himself. Then he called a referendum for approval of a new bicameral legislature and used his victory to emasculate the legislative branch of the government.

The country wasn't "ready" for a powerful legislature, Akaev insisted, arguing that Kyrgyzstan needed a strong presidency to institute economic and political reform. In referendum after referendum, he reshaped his office, winning the power to appoint all government ministers and ambassadors and to dissolve parliament. During the 2000 elections he triumphed in the polls although popular candidates had been jailed or disqualified, independent media shut down, and preprepared Akaev ballots found at one polling place. The weakened parliament became a hotbed of ineffectual opposition against a seemingly unstoppable president.

"So Akaev bears no blame, nor do the people who elected him?" I asked Tanya. "And precisely what do you want the United States to do? Meddle in Kyrgyz politics by removing him?"

"If not for the base—" she began, but the look on my face ended the conversation.

No more respectful of popular will than Akaev, the opposition was so busy dividing up their imagined spoils that they never went to the people to ask, point-blank: Do you really want to be ruled by a government that kills its own citizens? The government press turned the story into a tale of heroic Akaev struggling to control the unruly hordes. And the opposition press didn't even bother to compare the price of rubber bullets and tear gas to the cost of the metal jackets used on protesters in Aksy.

Several weeks later, Pyramid TV did air footage of the riot— shaky hand-held images of elderly women being beaten by the police, the crowd breaking and running at the crack of the first shots, the wounded writhing in pain, corpses lying motionless in the dust. But it provoked no outrage. I didn't even hear any consternation over the cameraman's admission that he'd turned the tape over to the TV station only after he'd offered it to the government in exchange for the governorship of the province and been rebuffed.

The only real disquiet I felt came from Mamasadyk, my young journalist friend. "This reminds me of 1999, just after the first Kyrgyz were killed by terrorists from the IMU," he said. "That happened just before Nooroz, when the president always addresses the people in Lenin Square. But he didn't even mention it. He just pretended everything was normal.

"That's what people want. *Normalna*. And as long as they can pretend that everything's normal, everything is fine."

15

Bishkek on the Potomac

An apple ripens by itself and falls into my mouth on its own.
—TATAR PROVERB

Nurjamal Asanova was an optimist.

In the United States that might not have made her particularly noteworthy, but in Kyrgyzstan, or at least in the Kyrgyzstan I'd come to know, that made her unique.

Throughout the long winter, I'd engaged in a solitary quest for the local version of a starry-eyed optimist, a dreamer with the capacity to surrender to the improbable, and Artyom, my favorite student, had assured me that I would fail. "It's not that we're pessimists," he'd explained, repeatedly. "We're realists. Americans, on the other hand, are unnaturally unrealistic, which is why you always seem sure that you can do anything."

Our cultures couldn't have conditioned us more differently. The music of Artyom's youth were the melancholy and brooding symphonies of Rachmaninoff and Tchaikovsky's dark melodies, what Harold Schonberg called "a particularly Russian kind of melody, plangent, introspective, often as emotional as a scream from a window on a dark night." My childhood had been infused with the syncopated rhythms and rhapsodic blues of George Gershwin.

Artyom had been raised on Pushkin, on Eugene Onegin's boredom, Tatiana's unrequited love, and Boris Godunov's uneasy conscience. The verses he'd memorized in school included lines like "Woe to a country, where the slave and flatterer both stand by the throne, and the bard, chosen by heaven, stands silent, his eyes downcast."

I, on the other hand, had been brought up on Walt Whitman, the bard of American democracy, to "a song of myself, a song of joys, a song of occupations, a song of prudence, a song of the answerer, a song of the broad-axe, a song of the rolling earth, a song of the universal."

Artyom struggled mightily against the natural pessimism of his culture, a constant foreboding of doom and disaster, and the only inspiration he'd found was, heartbreakingly, that piece of American preteen schlock, *Jonathan Livingston Seagull*, Richard Bach's fable about following your own path, even if your flock, or society, finds your individualism threatening. His penchant for Bach was the subject of considerable bemusement among his friends, who wanted to know, "What's the fascination with seagulls?"

Nurjamal, on the other hand, was a triumph of genes over culture. A young woman as striking, stately and diffident as Stephanie Seymour, the Victoria's Secret model, she oozed the self-assurance of the daughter of one of the country's most important families, gliding through the university corridors alone, friendly, yet never quite part of the gang. She smiled but never giggled, worried but never despaired, utterly confident that no problem could escape her resolve.

When she told me, casually, "My husband plans to be president of Kyrgyzstan one day," the pronouncement sounded entirely plausible.

In mid-April, Nurjamal approached my desk after class and said, more as expectation than as demand or request, "My husband would like to meet your husband."

Without seeking an explanation, I invited them to tea. Nurjamal had that effect on people.

"It is true that you're a businessman with an MBA?" Altyn asked Dennis once all the formalities had been completed. The formalities always included tea and some variation on the theme of "breaking of bread." That afternoon's "bread" was Dennis's specialty, homemade apple pie.

Once Dennis had established his theretofore-rumored bona fides, Altyn leaned forward in his chair, planted an elbow on his knee and continued, "What should Kyrgyzstan do about its water?"

Water was on everyone's mind in Kyrgyzstan since it seemed to be the country's only natural resource. Inventing an ingenious scheme for turning that resource into a cash commodity would be a career-making move for a young man on the make, and Altyn was surely on the make. He'd just finished a stint as an officer in the Kyrgyz military, "a necessary step for when he enters politics," Nurjamal had explained, and had taken a position in the Foreign Ministry.

In arid Central Asia, water meant the difference between economic stagnation and a booming cotton crop for an agricultural country like Uzbekistan or Kazakhstan. During Soviet times, no one considered measuring that water's value in rubles. It flowed according to the physics of socialist solidarity, as decreed by Moscow: water gushed down from Kyrgyzstan's snow-capped mountains and was collected in Soviet-built reservoirs, which were opened in the summer to irrigate the fields of Uzbekistan. The Uzbeks produced cotton that filled Moscow's coffers, some of which went back to the Kyrgyz as subsidies. Kyrgyzstan didn't need to tap its reservoirs in the winter for electricity since the exchange included a steady supply of Uzbek gas and coal.

But when the center crumpled, so did the deal. The Uzbeks realized they were sitting on a pretty valuable commodity and demanded that the Kyrgyz pay for fuel. The Kyrgyz, in turn, claimed that their water was as valuable as Uzbekistan's gas and that the two countries should continue to trade at parity. But possessing a somewhat firmer grasp of capitalism than the Kyrgyz, the Uzbeks understood that value was set by demand, not by some calculation of intrinsic importance, so they held fast in demanding cash. The

Kyrgyz could not or would not hand over much money, so each winter the Uzbeks shut off the gas, provoking no end of griping about the materialistic and avaricious Uzbeks.

The Kyrgyz would have loved to retaliate by shutting off the water, but they hadn't installed control valves, and in any event, they had no other buyers. At least I imagined that was why the Kyrgyz government didn't play tit-for-tat. They issued no explanation of their failure to engage in that game, which left the Kyrgyz people free to console themselves on frigid winter nights with the thought that they were too nice, too moral, too steeped in postsocialist brotherhood to deprive their former countrymen of such a life-giving necessity as water.

The situation was tricky because in 1992, filled with illusions about a new union of Central Asian nations that would serve as a mini-USSR, Kyrgyzstan had signed an agreement with Uzbekistan and Kazakhstan that established water management as a regional, rather than a national, concern and acknowledged that downstream nations like Uzbekistan and Kazakhstan had "equal rights" to its use. When a barter system was established, an exchange of oil and gas for 2.5 billion cubic meters of water from Kyrgyzstan, no mention had been made of cash payments.

But as Kyrgyzstan watched with growing envy the economic boom set off by oil in Kazakhstan and the cash flowing into Uzbekistan from its gas, Akaev began to complain: We're giving you all our water and getting nothing in exchange; we have so little water that we can barely irrigate our own fields; we don't have enough water left to generate hydroelectric power; we have to bear the financial burden of keeping up dams and water monitoring stations. Finally, in July 2001, he signed a new law that defined Kyrgyz water as a commodity with a price tag, the first clear warning that all previous agreements were about to be canceled.

"Water is like air," huffed the Uzbek minister of agriculture and water resources. "It cannot have an owner." A violation of international law, the Kazakhs proclaimed, threatening to turn Kyrgyzstan into a regional pariah. Look around the world: in more than one hundred water basins, you won't find a single example of charging for water.

Dennis knew all of this. You couldn't spend five days in Kyrgyzstan without hearing about the hottest political issue in the land. And while his executive MBA had taught him nothing about transnational water negotiations, it had schooled him well in playing for time.

"What about your water?" he asked. "The fact that it needs to be boiled and filtered to be safe to drink?"

It wasn't much of a diversion, but it gave him room to escape into the kitchen for more tea and a brief consultation with his wife. At least, that was the plan.

"No, no, our problem with Uzbekistan and Kazakhstan over water," protested Altyn, following him into the kitchen. "Moscow used to resolve this. Now Washington must."

I sat in the corner, listening, thinking: Right. Just what we need, to treat Kyrgyzstan as the fifty-first state. Especially since the arrival of U.S. troops, people around town had begun to joke about the prospect, calling Osh "Oshington," less a complaint, I suspected, than a poignant exercise in wishful thinking. The Kyrgyz had spent so long in the embrace of the Russian bear that they were feeling mighty cold in lonely independence, especially since they'd been carefully taught that they couldn't survive without a protective hug.

Altyn waited for Dennis to reveal some master plan by which Washington would turn Kyrgyzstan's water into a viable economy. Instead, Dennis suggested, "That's something you would have to take up with the embassy, not me, since I don't work for the government," hoping Altyn would grasp what few other Central Asians had, that most of the Americans they met had no relationship with the U.S. government.

Altyn was not deterred. "But America must do this. There is no one else."

Dennis moved on to plan B. "People seem to resent it when America intervenes in their affairs. If the U.S. worked out a compromise that you liked, the Uzbeks and Kazakhs would be angry with us, and if it was a compromise that the Uzbeks and the Kazakhs liked, you would be angry. So why would President Bush put the United States in that position?"

Altyn didn't even venture an answer, but I'd gleaned at least the emotional response to that question from my students: No matter the difficulty, the United States was obliged to embrace old Soviet responsibilities because it had emerged triumphant from the Cold War—a strange twist on the concept of the victor inheriting the spoils.

The arrival of the United States and allied troops had laid open the complex ambivalence toward the United States: America meddles too much in other countries, but America must help us; America must help us, but America shouldn't tell us what to do; America must help us, but not too much, and only in the "right" ways; America isn't serious about helping us if it wants anything in return; America must help us, but must not upset the balance of power.

I'd watched that ambivalence play out all year, on and off campus, in Kyrgyzstan and beyond. But every time I tried to reduce it to a single image, I returned to what happened when my university asked the U.S. Embassy for money to build a television studio. According to the KRSU application, that plan was the next step in the creation of a full program in broadcast journalism that had begun with the construction of two radio studios the British had bought for them.

I hadn't heard anything about a radio studio on campus, so I'd queried my students.

"It hasn't operated all year," they told me. "The university gave the room to the UN, and the equipment is sitting in boxes, most of it already stolen."

Puzzled, I then asked Katsev to explain the seeming contradiction between what the application stated and reality.

"Why is it any of the embassy's business whether we have a radio station? Why do they think they have the right to interfere?"

Altyn hung around our apartment all afternoon, drinking tea, eating apple pie, discussing the complexities of Kyrgyz-Uzbek-Kazakh water negotiations, the history of Soviet dam building and the politics of upstream-downstream water agreements in other parts of the world.

Surely, if Dennis understood the difficulties, he would wave his magic American wand and produce a solution that would halt Kyrgyzstan's descent into economic oblivion. Just as surely, Dennis reiterated that neither he nor the United States had the power to do so.

"It's not our place to intervene in a regional conflict," Dennis said for the fiftieth time. "Whenever we do, people criticize us for meddling."

Altyn finally conceded the point, but added wistfully, "But if America won't, who will save us, then?"

The sun was setting. Nurjamal was flagging from hours of translation.

Dennis knew that the Kyrgyz craved answers, not lectures on picking yourself up by the bootstraps. But he was too tired to resist candor.

"I think Kyrgyzstan needs to grow up and learn to resolve problems by itself," he said. "That's what it means to be an independent country."

Altyn gazed wearily out the window. Nurjamal leaned over and whispered, "But what if we don't know how?"

16

The Globalization Tango

He who holds on to two boats will drown.
—KYRGYZ PROVERB

Perched on tiny plywood platforms, two not-so-scantily clad Russian go-go dancers, their hair bleached into the consistency of straw, swayed indifferently to the oversynthesized techno beat behind a reverb-heavy mix of Martin Luther King delivering his "I Have a Dream" speech. Below them, outlined by incessantly strobing black lights, forty or fifty young people in Old Navy jeans, U-2 T-shirts and sweatshirts from the University of Nebraska or Cal State sat nervously at tables scattered around an utterly empty dance floor.

Emil, a Lebanese twenty-year-old who was the coolest kid at the American University, had organized the evening's entertainment, the first rave to be held in Bishkek. Young people knew from movies and music videos that raves were a cool American custom, but no one was entirely sure what a rave actually entailed, how they should dress or act. They waited expectantly, then, for the magic to descend.

I'd never been to a rave, although I suspected that something was off the minute I heard it was being held at a three-star hotel and that no nitrous-oxide-filled balloons would be available. But I couldn't pass up the chance to watch history being made on the new Silk Road.

Emil lounged by the door, collecting the 60-*som* entry price in low-slung pants and a ski cap—Beirut meets the ghetto, light on the Beirut. Emil's family had decided to wait out the Lebanese civil war in Nice, France, and had then moved on to Central Asia, so his roots in Lebanon barely grazed the sandy soil. His girlfriend, a sexy Russian with Bo Derek braids, hung on his arm, planning their summer vacation in Montreal in English, the only language they had in common. Inside the disco, a Korean girl and her best friend, a nineteen-year-old from Kashgar in western China, gossiped and giggled nervously in Russian, which neither could speak when they'd first met two years earlier.

Jerry, the campus Chinese DJ and rapper, buttoned and unbuttoned an old Beatles jacket, unsure which was the "right look." And Regina, my translator, fussed with her hair, envious of the fake tattoo the other Regina, the one from India, had penned onto her forehead, as a postmodern *bindi*.

Travelers are drawn to "happenings" like Russians to vodka or New Yorkers to trendy restaurants. Italians plan their summer vacations to America around the Hemingway Festival in Key West. The French drive to Clinton, Montana, for the Testicle Festival, where they sample the local cuisine, marinated and breaded tendergroin, as the locals call it. And Australians somehow find their way to Yellville, Arkansas, for the Miss Drumsticks competition and the annual dropping of terrified, squawking *Meleagris gallopavo* at the annual Turkey Trot.

Festivals, holidays and public celebrations hold out the promise of an unexpurgated glimpse of native folklore, a chance to mingle with "ordinary folk" outside the tourist triangle of hotels, restaurants and obligatory sites. You know, real Americans doing real American things, not cruising the mall or stopping at Starbucks on the way to work.

All my life, I've been drawn to that same tourist trail, finding there a unique window into other cultures. So I jumped over bon-fires in Ecuador on the feast day of Saint Peter and Saint Paul for a glimpse of the strange blend of Catholicism and indigenous reli-gions. I let myself be pounded with squishy tomatoes during La Tomatina in Bunol, Spain, in my quest to understand Spanish anar-chism. And in the Cotswolds, I joined the rest of the demented tourists, British and foreign, who rolled hunks of Gloucester cheese down Cooper's Hill, trying, yet again, to sort out British staidness from their propensity for slapstick.

The Kyrgyz treated holidays and celebrations with a unique zeal. Valentine's Day was a cause for congratulations—for what, I wasn't sure—from every stranger on the street. And International Women's Day, an old workers' holiday if there ever was one, was observed with parties, flowers and greeting cards, all without the heavy hand of Hallmark or FTD—and entirely bereft of dedica-tions to Alexandra Kollontai, the leading Bolshevik feminist seem-ingly having been forgotten by everyone but me.

Holiday or not, when you walked outside, you always ran into something. Shortly after we arrived, it was a bevy of ancient Soviet Army veterans decked out in ragtag old uniforms, their chests cov-ered with medals, declaiming à la Brezhnev when TV crews approached. We never found out what they were celebrating, or if celebration was the operative word. But President Akaev and his cronies were yukking it up in a special tent, carefully isolated from the horde of what we assumed were honorees by a phalanx of armed guards.

A few weeks later, we stumbled onto National Beer Day during a stroll through the amusement park in Panfilov Park, an acre of the same rickety rides you see in small-town carnivals. A line of beer stands had been set up along the wide avenues around it, and hun-dreds of drunken Bishkekites were prancing and dancing in kiosks set up with TVs, VCRs and microphones for karaoke-philes. Pho-tographers competed for business with faux fantasy settings: gaudy couches to convey a little romance, miniature cars for the kiddies and photo backdrops of palm trees for those hoping to convince relatives that they were vacationing on the shores of an actual sea, as opposed to Lake Issyk-Kul.

A coterie of ethnic Russians—the women in long, embroidered skirts, the men decked out as Cossacks—sang Russian folk songs in front of a parked truck that served as their makeshift dressing room, costume shop and prop box. The smoke from a dozen makeshift grills cooking fatty lamb burnished the edges off their sodden performance.

The real show, however, was unfolding on a stage set up in front of the colonnade of the parliament building, where party leaders had once surveyed the troops on May Day. It was a perfect patchwork of twenty-first-century life, beefy women belting out Kyrgyz folk songs and lithe young women lip-synching Courtney Love.

Then a group of seven well-endowed lovelies in miniskirts and form-fitting sweaters grabbed the limelight as Liza Minelli. The men pressed forward ever so slightly, panting at the preening beauties onstage. But to our eyes, something looked off.

"Don't they look like those guys in the dance act ten minutes ago?" Dennis asked.

Dennis was right. There was no bounce to the sweaters.

When life stopped being a cabaret, the dancers flung off their wigs with great fanfare and, in the most modern of moments, revealed themselves as men. Even the drunks froze in nervous silence, their mouths hanging open. We steeled ourselves for the jeers, for the anger of embarrassment. It never came. The women showgoers howled in approval. The men stood mute, too cowed or drunk to react.

A week before the rave, the same day that Donald Rumsfeld slipped into Bishkek almost unnoticed to rally the troops at the air base, I'd hiked over to the Sports Palace, an indoor stadium, for what became my clearest view of the odd waltz being danced by the Kyrgyz and the Russians. The occasion was a Kyrgyz-language competition, a cross between a slam and a competitive revue. The competitors were the team from KRSU, including Artyom, and the challengers from the Kyrgyz Technical University. Such contests always provoked fierce rivalries, but the stakes were high that afternoon. For the KRSU team, winning would strike a blow for Russian superiority: Look, we're not even ethnic Kyrgyz, but we will prevail, nonetheless, in the Kyrgyz language. For the students from

KTU, a predominantly Kyrgyz institution, losing a competition in their native tongue would have been the worst sort of humiliation.

More than 5,000 students packed into steeply banked bleachers, the amphitheater charged with the excitement of the Penn-Cornell game. The KRSU students waved mini-banners emblazoned with both the Russian and Kyrgyz flags, while the KTU fans lofted balloons with exclusively Kyrgyz symbols.

For four hours, male students dressed in the bizarrest drag I'd ever witnessed—the long, full skirts, embroidered blouses and befeathered hats of traditional Kyrgyz women—traded insults about presidents Putin and Akaev. Girls in miniskirts sang Kyrgyz translations of the Beatles, and the whole cast harmonized on tributes to the glories of their respective universities. At least, that's what I thought they were singing about. It was hard to hear, with the audience alternately booing and applauding, either deriding the other team—KRSU in black-and-gold ties, KTU in blue university T-shirts—or cheering on the good guys.

In one skit, six Russian babushkas, beggars swaddled in shapeless bundles of baggy clothes, mocked the current political crisis.

"The prices are too high," one said.

"Don't worry, the United States is planning to buy Kyrgyzstan," another reassured her.

"Then Talas will become Dallas." The audience roared.

Thankfully, it was then time for a faux commercial break. The product being sold: the Kyrgyz language.

"How do I get as tall as you?" a tiny KTU boy asked his companion, a veritable giant. The canned answer: "Learn your native language."

The KRSU team followed up with a catchy ditty that said, more or less, Even though we're Russian, we love the Kyrgyz language as much as we love our mother tongue.

Whipped up by their teachers, who served as resident cheerleaders, the KRSU students stomped, cheered and screamed, although in truth few understood anything being said onstage since few understood any Kyrgyz. Then again, none of the members of their team did either.

The contest was an elaborate, unacknowledged farce in which

students like Artyom, whose Kyrgyz was limited to the equivalent of "hello," "good-bye" and "I don't understand Kyrgyz," wrote their scripts in Russian, then turned them over to a Kyrgyz teacher for translation. The endless show, then, was an exercise in patriotic fake fluency performed by young people who already had made plans to leave the country.

"What was the point?" I asked Artyom, whose team had emerged triumphant.

"Most of the students were required by their teachers to participate. And we all thought it would help us with our grades."

But the scene at the Dostuk Hotel the night of the rave was both more authentic and, perhaps for that naive authenticity, more bizarre than the self-mocking Testicle Festival, La Tomatina or the Kyrgyz competition. China and India traded music, fashion and culture with Europe and America in a modern remake of the cultural mix of the old Silk Road.

Two years earlier, the Dostuk had been the scene of a spectacular murder of Chinese Security Ministry officials by members of the Uighur Liberation Organization, the most militant of the Muslim separatist factions from western China. That night, rap and hip-hop wafted out of the hotel disco onto Victory Square, which was dominated by a huge statue of Soviet troops defending the motherland from Nazi infidels, although, as usual, the "eternal flame" was unlit.

When Kyrgyzstan was still ruled from Moscow, the Dostuk had been Bishkek's premier hotel, the hostelry of choice for the poobahs—the "people's representatives"—from the Kremlin. With 179 nearly empty rooms, the decor had achieved a state of neglected shabbiness that screamed of an interior decorator with faded glory in mind. In the disco, at least that theme worked as the "industrial look," especially under the black lights.

The few American University students who were actually American bought Russian beers in English, Turks and Turkmens purchased Coca-Cola in Russian, and Jerry, the Chinese rapper, regaled anyone who would listen with tales about the glories of his hometown, Kunming, down by the Vietnamese border. "It's a city of eternal spring," he gushed. A third-generation Chinese Christian majoring in business administration, Jerry had moved to Kyrgyzstan to perfect his English so that he could enter a Christian seminary in Iowa. "And of course, as you will recall, it was the site of the International Horticulture Exhibition in 1999."

No one remembered that exhibition, nor was anyone much interested. The beat of the music was beginning to overcome the shyness. Joy, whose real name was Gulzada, was trying to pull a group of girlfriends onto the dance floor. Her time in the United States had been spent at a Christian college, but Joy had a punkish tendency to dye her hair a different color each day. That night, she was wearing grunge copied almost perfectly from the pages of *Seventeen*. But with her butch haircut, the look screamed "young dyke with an attitude," although no one in the room was savvy enough about America to hear that message.

Regina, my translator, was too fixated on the go-go dancers to join her on the dance floor. "They're prostitutes?" she asked, horrified at the public display. The curriculum at her suburban Dallas high school apparently hadn't included any lessons about go-go dancing.

The AUK students were members of a transnational lost generation unmoored from any single culture or nationality. Iranian by birth, Ahoura Afshar was a Bahai refugee from a homeland where Bahai students were denied admission to the university, sacred properties were confiscated, if not destroyed, and Bahai marriages were not recognized by the state, turning Bahai children into bastards. His parents lived in Kazakhstan, his sister was studying in Budapest, and he was waiting to hear about graduate school fellowships in the United States and England.

The daughter of the head of Central Asian programs for the Soros Foundation's Civic Education Project, Jessica Buckingham was Canadian, with a boyfriend who was half Uzbek, half Uighur. Regina Matharu was Indian, but born in Moscow, raised in New Delhi, and absolutely trilingual at the age of nineteen.

There were Kazakhs who were more American than Central Asian, Mongolians indistinguishable from Kyrgyz, and Chinese so confused by what they were that they figured they might as well move to America, where "no one is really anything, or maybe they're a little of everything," as one student put it.

Many of AUK's students were alumni of U.S. government exchange programs designed to promote "cross-cultural under-standing," a code phrase for winning the hearts and minds of young people in potentially hostile nations. The program had both worked perfectly and backfired absolutely: the kids had been so thoroughly won over that they were all angling for ways to move to the United States instead of plotting their careers as future—and pro-American—leaders of their countries.

AUK's clash of cultures and assumptions reflected every catch-22 in meshing tradition with modernity. Religious Muslims wouldn't attend classes on Fridays, religious Jews couldn't study on Saturdays, but classes were not in session on Sundays, although there were few Christians on campus. The cafeteria served local food, which strict Muslims wouldn't eat because it contained meat not butchered according to Shari'ah guidelines, and every year there was a hassle to change the cafeteria hours during Ramadan to accommodate kids who were fasting.

That spring, the firing of a British faculty member had erupted into a tussle over cheating, academic standards, grading, homosexuality and what exactly the phrase "American University" meant in traditional Central Asia.

The first version I heard of the conflict was that Barrie Hebb, a Canadian teaching economics at AUK, had caught several students cheating and, assuming that working at the American University meant that he should hold his students to Western standards, flunked them, breaking rule number one: Thou Shalt Not Flunk Students. That wasn't AUK's rule; it was Kyrgyz tradition. No student failed. If his exam grade was too low, he retook it. And retook it. And retook it until he achieved a passing grade. AUK students were proud—strike that, unabashedly arrogant, even drippingly supercilious—about the superiority of the most expensive higher education in Bishkek. But they weren't about to do any work to earn that cockiness.

They were outraged, then, at Barrie's breach of etiquette. "He should have been fired," Regina told me. "His job is to pass students, not flunk them."

Their parents, the local version of Ivy League mothers and

fathers, were even more incensed and delivered their fury directly to the president's office. The president, an American named David Huwiler, had tried to "reason" with Barrie, who acted like a Western faculty member and told the president to stuff it. When he was fired, a small group wondered if Barry had really been fired over his grading or if the president had another agenda.

That's where version two came in. Barrie had been involved with a student, a not uncommon occurrence in a country where the phrase "sexual harassment suit" had never been uttered. None of the other faculty members who'd slept with, married or impregnated students had been called on the carpet. Barrie, however, had gotten involved with a student who was male, not female, and both he and his boyfriend, Konstantin Sudakov, a Kazakh so Americanized that he would have fit right in at a Queer Nation demonstration, smelled a rat.

According to Barrie, Huwiler didn't fire him directly. He didn't write him, e-mail him or call him into the office. He just told the director of the Civic Education Project, which paid Barrie's salary, that the young Canadian was no longer welcome on campus. In the course of their conversation, however, Huwiler allegedly mentioned, in passing, Barrie's relationship with a male student as potential grounds for dismissal.

A valuable instructor who could teach in both Russian and English, Barrie was transferred to another university in Bishkek, and the matter might have died without much attention if Konstantin hadn't decided to act like a journalist, an American journalist. Another alumnus of a U.S. exchange program, Konstantin wore his American self with such easy familiarity that it was an effort to remember that he was from Karaganda, not Kalamazoo. During his year in Virginia, Konstantin had discovered that he didn't have to suffer fools, gladly or in sorrow, and he'd carried that confidence into his work with the AUK student newspaper.

After Barrie's dismissal, Konstantin had appeared at the AUK president's regularly scheduled open office hours and questioned Huwiler, with the same edge Sam Donaldson might have brought to an interview with Hillary, about AUK's hiring—and, by extension, firing—practices. He didn't mention Barrie's name, but he grilled Huwiler about the university's habit of recruiting faculty through the old-boys club system. "What about equal opportunity employment? Affirmative action?"

Huwiler had taken to Kyrgyzstan, at least to its authoritarian nature, with the same aplomb with which Konstantin had shaken off Kazakh passivity. He was not amused, then, by Konstantin's in-your-face American demeanor and fired off a letter accusing him of being a journalistic hack grinding his own ax because his boyfriend had been fired, a letter he invited—nay, dared—Konstantin to print in the student newspaper. Konstantin called his bluff, leaving Huwiler with at least a crate of eggs dripping from his face. But Huwiler, still not quite getting it, sent Barrie an e-mail accusing him of conspiring with the student newspaper to undermine his position as president even as he vowed his support for gay rights.

Rumors about the first openly gay student and his relationship with a teacher had electrified the campus, and the AUK virtual forum exploded in homophobia. "Do you really think that gays are people who deserve to be treated as people?" one student wrote. "It's against nature. Line 'em up and shoot them." Another seconded that sentiment, "If I had a SON who was a GAY I would execute him right in Ala-Too Square."

In the early hours of the rave, the Konstantin-Barrie gossip coursed through the disco to the titters and uncomfortable guffaws of teenagers who'd never met an uncloseted homosexual. The rumored details—"Is it true that they 'did it' in his *office?*"—and the breathless speculation—"Do you think there are *others?*"—provide a welcome relief from the tension of the empty yet beckoning dance floor. They might have been the hippest kids in Bishkek, but neither high nor drunk, they turned the rave into a junior high cotillion. Girls danced with girls, boys with boys, and potential couples eyed one another from across the room, too shy to make the first move. The scattering of students from other universities—easy to spot since their hair and makeup were perfect—hung back, even less bold, by the bar at the far end of the room.

"What makes this a rave?" I asked Regina. No drugs, little alcohol and none of the bruising anger of would-be punks in the United States, who in comparison had comparatively little to be angry about. In one of those strange generational and spatial anomalies, in Kyrgyzstan it was the old, not the young, who acted out.

"I don't know, that's just what we call it because, you know, it's an American thing," she replied, playing with a dozen plastic bracelets, which were glowing in the dark.

"Come on, let's dance," yelled Dennis, trying to make himself heard over the 5,000-decibel music. Ten minutes later, Duran Duran's "I Wanna Take You Higher" wound down, and we collapsed back into our seats.

"Did you get high?" Jerry, the Chinese rapper, asked. I sensed no irony in the question. "I *love* these American customs!"

part FOUR

17

Guess What's Coming for Dinner

When you arrive at my home is up to you.
When you leave is up to me.
—KYRGYZ PROVERB

To my mind, at least the part of my mind forged as a college professor, May is the climax, the moment of the final sweeping lectures
that hammer home the big picture, the chance to reflect on a job
well done, or on the roots of abysmal failure. But nothing at
KRSU meshed with my mind, and in Kyrgyzstan, May was canceled. Sure, May—Mai, in Russian—appeared on all the local calendars, but it had been pretty well erased from the work and
school schedules. May 1 was International Workers' Day, May 5
Constitution Day, May 9 Victory Day and May 29 Armed Forces
Day. In between, most people would be so busy recovering from
celebratory drinking, preparing for celebratory drinking or engaging in celebratory drinking that the country would come to a
standstill.

My students elucidated this reality in inordinate detail on the same afternoon that Artyom asked me to explain why the Kyrgyz standard of living was so much lower than the American.

I was gearing up for a serious discussion of mineral wealth, the challenges of new independence and historical baggage when Dinara chimed in.

"Americans work harder," she said bluntly, describing her shock when she realized that her American host parents regularly stayed at work until 7 P.M., caught up on reports at home in the evenings and stopped by their offices on weekends.

"That shows the way capitalism is," one student chimed in. "Workers are treated as slaves."

How was I going to explain this one?

"It's a little more complicated than that," I began, talking about how many hours I worked, even without a boss.

"But why do you work that hard if you're not forced to?" they wanted to know, pushing me into a long explanation of the complicated gallimaufry of pride, ambition, love for my work and desire for a comfortable life that drove me and most Americans I know.

"So you do agree that we don't work so hard as Americans?" the students pressed.

Backed into a corner, I couldn't resist running down the list of how little the Kyrgyz labored: forty-five days off after any death in the extended family, four or more weeks of vacation, six months' paid maternity leave, retirement at fifty-five for women, sixty for men, as well as at least three Muslim holidays, Orthodox Christmas, Independence Day, Commemoration Day, International Women's Day and the entire month of May. I didn't mention the amount of time workers spent drinking tea. That seemed like hitting below the belt. After all, during the Soviet era, back when it made no difference to your salary whether you did any work at all, workers were guaranteed two full months of paid vacation.

Communism might have fallen in Kyrgyzstan, but old habits hadn't changed.

Since it had taken me a full semester to figure out enough about my students to begin teaching them, I was seething with frustration at the cancellation of May. Dennis and I were about to leave for three weeks in Iraq, the first issue of our newspaper still hadn't appeared, and I knew I'd feel like a failure if we didn't get at least two published before the end of the year. But there was nothing to do except distract myself. A road trip, we decided. Let's take our battered 1978 Mercedes out for a drive through the hills.

Road trips were an unknown luxury in Kyrgyzstan, a country with few functional vehicles and even fewer functional roads. Even foreigners accustomed to driving across the California desert or the European Alps resorted to taxis, buses and the occasional private driver rather than purchase cars—out of a generalized sense that "the roads are awful and the people drive like maniacs." We accepted that perplexing reality until November when we realized that back home we lived on a dirt road in New York State and were accustomed both to awful roads and maniacal drivers.

All we needed to buy a car was a Russian-speaker with a wily instinct for negotiation. Samarbek was both, and he offered us a deal we couldn't resist. Knowing that we'd been thinking about buying a yurt, he proposed that his brothers-in-law, both mechanics, help us purchase a vehicle which we would give him at the end of the year in exchange for a furnished yurt that his mother would make.

So on a lovely Saturday morning in the fall, Dennis and I, Samarbek and his brothers-in-law drove down to the Vorobiev Bazaar, a huge field peppered with rusting Ladas and dented Toyota Corollas, shiny BMWs, vintage Volgas and stolen Mercedes driven in from Poland and Lithuania by the dozens. Most of the sellers snoozed in their cars, waiting for someone to show up. Some had been napping there for weeks.

We dismissed the Ladas out of hand, having learned that Russia's idea of a reliable vehicle was one that started most of the time, didn't lose its window handles for at least six months and needed to be fixed only once a week. The old Volgas tugged at all my fantasies about classic vehicles, dreams I thought I'd laid to rest when I drove a 1957 Chevy across the United States. Dennis restrained me. Just what we needed, a classic Russian junker!

Finally, Samarbek's brothers-in-law sealed a good deal—$1,500—for a used Mercedes, and I was ready to zoom off in delight at the ease of the transaction. Little did I know that the work had barely begun.

Registering a car in Bishkek was either an exercise in the type of patience few Americans can muster or an all-day experience in bribery. Fortunately, Samarbek's brothers-in-law were experts at greasing the palms of the long line of bureaucrats who stood between a car owner and his ownership papers, so we opted for the latter. For just $6, three-hour waits miraculously turned into five-minute perusals. Overlooking a missing signature cost a mere $3, and deeming the threadbare tires roadworthy was a bargain at a buck.

I don't want to leave the impression that navigating through the Soviet-style bureaucracy was easy, even with a ready wad of 200-*som* bills. One stamp was never enough, so just when you thought you'd finished up in the registration office, you were sent down the hall to another cubicle where a grumpy, attitude-heavy woman moving at the speed of a 300-pound turtle examined your documents yet again before turning them over to another government worker with a grudge against work for a different stamp. She, of course, couldn't complete the final product until you stood in another line and bribed another person for the privilege of paying for a stamp that certified that you had paid for the first stamp. With that receipt in hand, you were sent to another building, miles away, for still another stamp, which you then had to take back to the first building in order to get your precious plates.

"That's it, right?" Dennis asked once he had his plates in hand. By then, he'd been at it for four hours.

"No, now we need inspection," Samarbek's brothers-in-law explained, a tad impatient with Dennis's impatience.

The inspection station was a dirt road—not *on* a dirt road, but the dirt road itself—in front of a shack, where, for a couple of dollars, an official was only too happy to put down his tea, glance at the car and send it along to another office in the far suburbs. Unfortunately, that office was manned by no one in the network of friends, relatives, neighbors and friends of relatives that had smoothed the way thus far. Unwilling to tempt fate, the brothers-in-law drove around until they found a vodka-infused neighbor who knew someone who had a cousin at the inspection station.

Unfortunately, that close contact had taken a long tea break. Left with no alternative, Samarbek's brothers-in-law called a break of their own, a vodka break. Finally, two hours, three vodkas and $6 later, the inspection deed was done.

"I can't believe that it took from 8 A.M. to 5 P.M. to get our car registered!" I complained to my students the next morning.

"You did it in one day?" they responded, in amazed admiration. Those wily Americans! "That's incredible!"

By the time we purchased the car, winter had already descended on a mountainous country without a single snowplow. May, then, was the first chance we'd had to take to the open road.

For months, Azamat—the student who'd befriended Dennis at the Radio OK English class—had been begging us to go home with him to the village of Chayek, and Dennis had always teased him, "Fine, but I don't eat scrambled goat brains."

Azamat didn't really get it. Who wouldn't eat scrambled goat brains?

"It's just that in our country, we're not accustomed to eating goat brains, and the thought of it is, well, slightly repulsive."

It wasn't just the brains. According to Kyrgyz tradition, guests merit an entire goat, slaughtered in their honor. After it is cooked, the head is ceremoniously placed before them. They are expected to pop out and munch on the eyes, then crack open the head, scoop out the brains and share them with everyone else at the table.

In my younger days, even monkey, snake and guinea pig hadn't stopped me. In my older days, I agreed with Dennis: one of the advantages of age is that you don't have to eat anything that has recently been your playmate.

We'd heard all the goat details, with a full description of the texture of the eyeballs and the difficulty of cracking the skull, from our friend Eric, who ran the local Habitat for Humanity office. "It was sickening," he'd told us, his body shuddering at the description. "But I had to pretend that it wasn't because I didn't want to insult my hosts."

Goat brains weighed heavily on our minds when we agreed to turn our road trip into a visit to Azamat's family—on the condition that he allow us to turn the tables by treating him to a stay at the Aurora, Kyrgyzstan's fanciest resort. A journey to the two poles of Kyrgyz life—like a stay at the Ritz combined with a stopover at the home of a sharecropper in rural Mississippi—seemed a pretty convincing distraction from my travails at KRSU.

Intent on avoiding any insult, we'd pressed Azamat mercilessly to tell his parents not to kill a goat in our honor.

"But it's our tradition," said Azamat.

"But it's not mine," replied Dennis.

"It will be a shame on my family if they don't slaughter a goat," Azamat persisted.

"I won't consider it to be shameful," Dennis countered.

"But the whole village will notice, and they will," Azamat begged.

Suspecting that as citizens of a culture with somewhat less rigorous traditions, Dennis and I were bound to lose this cultural conflict, I'd suggested that Azamat offer his parents an out, an explanation they might accept: "Tell them that we're vegetarians, or that our religions do not permit us to eat goat."

No one smiled quite like Azamat, whose face lit up with his grin, broadcasting unmistakably perfect childlike wonderment. He did not smile at my proposed compromise. "I'll try," was the best he could manage.

Azamat skillfully evaded the goat-brain conundrum as we took to the open road, our hearts humming with that unique sense of freedom that courses through American veins the minute we touch our hands to the steering wheel and press down heavily on the gas pedal.

In a country that's in the middle of nowhere, Chayek was at least 150 miles from anywhere. Even accounting for a certain amount of circling and swooping, crows probably could have flown there in three or four hours and human beings had once been almost that speedy. But the short road to Chayek hadn't been maintained since the hammer and sickle stopped flying. Villagers traveling back and forth to Bishkek were forced to drive halfway to China—literally—then cut back toward the center of the country along a road missing whole sections of pavement, leaving potholes that consumed full wheels.

Mile after mile, the road wound through green rolling hills, up snow-capped mountain passes and across fertile river valleys into the heart of traditional Kyrgyz life. We saw not a single plowed field, not a single farm. Even seven decades after collectivization, the Kyrgyz still think of themselves as nomadic shepherds.

Nature gave the Kyrgyz a stunning landscape, but human beings had succeeded in marring much of the beauty. Trash collected in heaps by the side of the road. The rust of abandoned hulks of Ladas and ancient trucks leached into the soil. Abandoned and semiabandoned coal and uranium mines were left unguarded, to be picked over by locals unaware of the risks. The only thing missing were billboards, a modern invention that had not yet blighted rural Kyrgyzstan.

Under communism, Chayek had bustled with the energy of producing meat and potatoes for the Soviet people, with a constant stream of villagers from surrounding communities, and with a rich mix of Russians, Tatars and a few Germans living side by side with the Kyrgyz in proletarian harmony. No one had indoor plumbing, but electrical and telephones lines had been installed. Most importantly, everyone had a job.

But the Chayek we saw was a silent village. There were no cars on the road, and no farm equipment belched through the stillness. The only gasoline available was dispensed by the bucket.

As former members of a single collective farm, each family had received a plot of land when the collective was dissolved, but few could make a go of it. The old managers had grabbed the most fertile properties and all of the equipment. But they, too, were having a tough time trying to figure out how to develop a market, or calculate profit. With no work to be found, the community had devolved into near-subsistence farming.

Azamat's father raised a few animals and maintained a small

garden, but he was neither herder nor farmer. Under communism, he'd worked as the collective farm's mechanic, keeping trac-
tors and mowers in repair. These days, there wasn't a tractor worth
repairing within a hundred miles. The only regular income the
family received was the token salary brought in by Azamat's
brother, an assistant to a bus driver, which they used to buy sugar,
rice, matches and salt. They baked their own bread using grain
that they grew, drank milk from their goats and mares, and
bartered potatoes from their garden for coal to heat their home in
the winter.

The men of the village spent most of their time wandering from
house to house, drinking shots of homemade vodka, visiting, and
instructing their wives to prepare tea laced with mare's milk. The
women complied, and worried.

The Russians, Tatars and Germans were all gone, part of the
huge wave of Kyrgyz citizens who'd fled when the country became
independent. The village schools had been left with little money
and few qualified teachers, so only a handful of students were
learning much Russian. The older generation wasn't complaining,
unwilling or unable to discern the implications of emerging mono-
lingual in Kyrgyz, a language spoken by fewer than 5 million people, in a global world.

"Unless the Soviet Union comes back, or America adopts us,
what else do people have left but the past, but tradition?" Azamat
asked rhetorically.

But the pride the Kyrgyz avowed in their traditions was at
strange odds with their humor, with the stories that they delighted
in telling about themselves.

*While walking in the countryside, two Uzbeks and two Kyrgyz fell
in a hole. "I'll give you a hand up," the younger Uzbek said to the
older. "Then, when you're on solid ground, you can pull me up." The
older man agreed, the Uzbeks freed themselves and went on their
way. The two Kyrgyz looked at each other grimly, and one began
climbing out of the hole on his own. "Hey, you can't do that," yelled
the other man, pulling on his companion's legs. "If you get out, I'll be
alone and stranded."*

Try another:

The devil captured two Uzbeks and two Kyrgyz and decided to make a stew of them. He filled a pot with water, placed it on the fire, lowered the Kyrgyz inside and covered them with a lid. Then he filled another pot with water, placed it on the fire, added the Uzbeks, and covered it with a lid, and then he nailed the lid down tight.

"Why don't you nail our lid shut?" yelled the Kyrgyz from inside their pot.

"It's not necessary. You're not like the Uzbeks. You won't help each other get out."

I could say that I'm an inveterate collector, but that's an evasion, a way to make my addiction to shopping sound high-minded. Heir to my father's predilection for what others consider to be garbage—his favorite shopping expedition was a weekly excursion from his home in Jerusalem to the Bedouin market outside Beersheba in southern Israel—I gravitate not to malls or trendy boutiques, but to third-world bazaars, weekly village markets and the kind of antique stores most people think of as junk shops. Dennis is my perfect partner in crime.

Our apartment in Bishkek was already overflowing with old carpets from Kyrgyzstan, wedding hats from Afghanistan and handmade felt rugs from Turkmenistan. We'd filled boxes with traditional robes, camel saddle bags, earrings, necklaces, handmade toothpicks, boots, scarves and yurt decorations. Dennis had bought 25 *kolpaks* as gifts for all our friends, for business acquaintances, even for the members of our local town council. After we found a full set of Ayatollah Khomeini plates in Iran—genuine Melmac—we fixated on throwing a dictator dinner and had purchased Turkmenbashi tea and watches, a Shah of Iran ashtray and Mao Tse-Tung vases for the table. We craved Soviet kitsch to lend our dinner a more "proletarian" character.

Azamat laughed at the prospect of American capitalists turning icons of communism into souvenirs, but he was sure he could help us collect it. "Everyone in my village has old pictures of Lenin, busts of Stalin and banners from May Day parades," he said. Since everyone in his village was broke, he assumed they'd all be willing to part with this now-irrelevant memorabilia.

Azamat was wrong.

No statues or paintings of the Kyrgyz mythic superhero Manas adorned their homes, but old Soviet heroes were too sacred to be disposed of, even for the equivalent of an average monthly wage. We managed to buy two small busts of Lenin—Lenin as Rodin's Thinker and Lenin's glowering face—but Stalin? No way. Not a single family was willing to part with old Joe, the national secret pleasure.

Stalin had been banished from public, from parks, university campuses and office buildings. Yet when we stopped at a tiny shack that served as the office and garage of the tire repair shop in Kochkor, on the way to Azamat's village, Uncle Joe still claimed a place of pride, right next to Lenin. And at the home of one of Azamat's relatives in Chayek, a Party official turned mullah, Azamat pointed to the corner that had long served as an altar to old Joe. "He keeps his Stalin photographs in the back room," Azamat whispered. "Just in case."

Even as they took refuge in the distant past, the Kyrgyz kept one foot planted in more recent glories. Traditions aren't etched in stone, after all, and despite the grim predictions of imminent fundamentalist uprisings, it was communism, not Islam, that still laid full claim to Kyrgyz hearts.

Within an hour of our arrival at Azamat's house, his father and the other villagers who'd dropped by to see the unlikely American guests were pressing alcohol on me.

"No thank you," I declined, trying to disguise my puzzlement.

I don't drink, a bit of abstinence that was inexplicable in Bishkek, where the Russian tradition of serving up vodka on every occasion and nonoccasion held sway. The government seemed to encourage the rampant consumption of alcohol that left the streets littered with sprawling drunks, lowering the price at any inkling of popular discontent. After the Aksy riots, a liter of vodka could be had for 75 cents.

But Chayek was a Muslim village, so I was caught unprepared. When I recovered, I mumbled, "It's against my religion," suspecting that "I don't like the taste of alcohol" would not be persuasive.

"God will understand," replied Azamat's father, throwing back a water glass of the clear brew. No one else at the table abstained, just as no one else at the table abstained from pork or went to the mosque despite frequent protestations that "Islam is part of our tradition."

The intermingling of traditions is rarely an easy marriage, but that theoretical dissonance didn't seem to bother anyone but me. "We've been Muslims for hundreds of years—it's our identity, and we were Communists for only seven decades," one student explained. "But Islam didn't guarantee jobs and electricity."

Dinner was served promptly at seven in the dining room where we'd already been served two other meals since our arrival in Chayek at 3 P.M. We'd steeled ourselves for that moment all day, ever since we'd learned that Azamat had been unable to grant us our only, beseechingly delivered, request.

We'd begun to suspect that Azamat had failed us when we heard the final bleats of a pretty young black-and-white goat that we'd petted just moments earlier. And our suspicions were confirmed when we returned from our stroll through Chayek's block-long downtown to find Azamat's entire family gathered in the courtyard, his father cleaning blood off a long knife, his brother hacking meat into hunks, and his sisters squeezing out yards of intestines. A massive pot of boiling water steamed a beige froth that sizzled onto the wood fire. Bones jutted out of flesh that bobbed next to a boiling head, its lips curled, revealing a macabre sneer.

It seemed that Azamat hadn't tried to talk his parents out of observing the time-honored Kyrgyz tradition. He simply couldn't. Hardly a sophisticate, he'd hung around with enough foreigners to understand that they had strange ways. But his parents didn't know anyone who wasn't from Kyrgyzstan. His mother, born in the tiny village of Communism, didn't even speak Russian.

The smell of the delicacy pervaded the small house. I was

already gagging when I took Azamat aside and whispered, "I just can't do it."

Smiling sweetly, he led me into a room piled with skins and quilts and handmade carpets that could serve as a bed, then apologized to his family, "Elinor is very tired."

Hanging tough, Dennis took his seat at the table, already groaning with nuts, dried fruit, sour cream, bread, jam and other delicacies that, blessedly, had no mothers. Then, beaming with pride, Azamat's mother entered the room and placed before him a heaping platter of boiled goat parts, the head set majestically atop the meat and organs, the lungs bulging with what looked like a milky pudding.

Ashen-faced, Dennis raced out of the room and joined me in my cocoon of rugs and quilts.

The next morning, I was dragged down with shame that we had refused the hospitality of our hosts. Azamat had been the best of friends. When we needed a translator, he never failed us. When the heat went off and I was shivering, he brought me his own electric heater, the only source of warmth in the tiny, two-room house without plumbing that he called home. I was mortified at the thought that I'd embarrassed him with his own family.

Azamat shared that sentiment. "Please don't feel bad about what happened. It's also my fault because I couldn't bring myself to tell my parents not to slaughter a goat for you. We think of this as hospitality, but I realize now that it has nothing to do with hospitality because it has nothing to do with the guests and their wants. People are just worried about what the neighbors will say.

"Tradition is a problem. It leaves us no room for us to develop ourselves. It is like a rope around our necks."

I assume that, by now, you're pretty sick of the word *tradition* and might well be suspecting that I'm overplaying my hand, or that no editor read this manuscript carefully enough to clue me in to the fact that I've repeated myself. But the word *tradition* was as fundamental to the vocabulary of Kyrgyz life as, say, *local control* was to the Gingrich revolutionaries. And since it's the elephant in the closet feeding a dozen international angers, tradition demands a short digression.

I arrived in Central Asia infused with a full measure of starry-eyed illusions about "traditional societies," which, in places like Manhattan, Los Angeles, Rome or London are said to be more genuine, more authentic, than the rituals of getting drunk on Super Bowl Sunday, the brawls at working-class London soccer matches or riding Walt's Magic Mountain. I admit that I'm not sure what the adjective *authentic* means when used to modify a noun like *society*, since it suggests that those of us who reside in "inauthentic" ones are characters on *The Truman Show* or black-and-white citizens of Pleasantville.

Curiously—or is it ironically? or tragically?—the word authentic crops up most frequently among tourists in reference to places with high concentrations of colorfully dressed villagers who live in poorly heated houses under handmade—a buzz word—thatched roofs and farm wildly beautiful but worn-out land with beasts of burden that include not only oxen but their own adorable children.

All of which leaves poverty and stagnation inexorably bound up with the Western definition of the words *authentic, earthy,* and *genuine.*

Embarrassed by my own penchant for the politically incorrect, I thought about none of this when I fantasized about the ingenuity of Kyrgyz yurts, the cleverness of Kyrgyz folk remedies or the exotic taste of fermented mare's milk from the safety of my home in New York. But I discovered that I make a lousy romantic. After nine months, I couldn't stop thinking about how cold those yurts were in the winter and how many infants and elderly died from respiratory infections; that that unpasteurized milk was serving up a lethal brew of Salmonella, *E. coli*, brucella, toxoplasmosis, cryptosporidium, tuberculosis, and staph; and that their folk medicine was powerless against cancer or automobile accidents.

Tradition wasn't just holding people back—a value judgment even I made only hesitantly. It was killing them, literally, having become an excuse for every possible exercise in corruption, authoritarianism, and bad behavior from bribery to lethargy, xenophobia, powerlessness, homophobia, wife beating, the government murder of its citizens and the worst newspapers I'd ever read.

No relief could be offered to women whose only way out of abusive homes was to douse themselves with kerosene, and nothing could be done about the corruption that diverted some absurd proportion of the nation's wealth into luxury jaunts abroad for the denizens of the *Beli Dom,* the White House, because those realities were foreordained by a near-biological imperative called tradition. Young women underwent hymen replacement surgery, young men were afraid to tell their families that they had converted to Christianity, women got kidnapped, newspapers couldn't tell the people what they needed to know, and no one could even discuss those realities because speaking of unpleasantries violated tradition.

The quandary wasn't unique to Kyrgyzstan, of course. In the United States, tradition had long been the defense for slavery and segregation. In Kenya, it kept the practice of female genital mutilation alive. In Germany, or Rwanda, it was the excuse for the wholesale slaughter of those defined as "other."

But in Kyrgyzstan, people were so busy serving tradition that no one ever thought to ask—because it wasn't in their tradition to do so—whether tradition was serving them, either as individuals or as a country, or whether it might need some tweaking. Tradition is continuity, a port in a tempestuous world, a sense of "we" when "I" can't quite cut it. But it isn't a blanket that descends whole cloth over a society. It's always selective, a winnowing of a mass of rites and rituals and beliefs that accumulates over the course of centuries.

I tried to say some of this to my students, arguing that forcing people to serve tradition was getting things backward and that tradition, like government or clothing, could be changed.

"We once had a tradition called slavery," I blurted out one morning. "It was a bad tradition, so we got rid of it."

As usual, their eyes glazed over, but that day I banned the use of the word *tradition* in my classroom.

If I had learned one thing during my year of travels, it was that tradition wasn't just quaintly benign. Nor are its dangers limited to those it enshrouds. When threatened, tradition fights back. It provokes annual anti–Valentine's Day riots in India, where Hindu extremists campaign against the growing influence of Western culture. It sets off rampages in southwestern Nigeria at the prospect of a visit by a German Christian evangelist. And it sends its defenders to topple its antagonists, as the whole world discovered on September 11.

It was only 10 P.M. but the lobby of the Aurora was as lively as the foyer of an old-age home at 2 A.M. The snack and gift shops were closed, the other guests had retired after a strenuous day of stomach lavage and electrotherapy, and the desk clerks had gone home for the night, leaving two security guards as the only "officials on duty."

But I was freezing, my usual state since the heat had been turned off, and I figured there had to be a way to extract an extra blanket from the resort that still drew the crème de la crème of Russia and Kazakhstan.

Named in honor of the battleship that fired the first shot of the Russian Revolution in October 1917, the Aurora had long been the destination of choice for party cadres from across the Soviet Union and a highly coveted vacation award for exemplary workers. None of the comrades seemed to care that the hotel and spa sat on a lake that was home to a Soviet weapons testing facility, an installation to recover uranium from the lake waters and a major uranium tailings site. Or perhaps that was part of the attraction for customers who often chose the spa at Jety-Oguz over the Aurora because the former was famous for curative waters with a high content of radon.

We'd arrived at Lake Issyk-Kul on a sparkling afternoon when the snow-capped peaks reflected in the placid turquoise waters lapping at miles of deserted sand beach. Since Issyk-Kul is one of the deepest lakes in the world, its waters are intensely blue. They should also be pristine—and they would be if villagers living along the lake's shores hadn't used it as a dump.

Scores of hotels and spas lined the north shore of the lake, but no facility had greater panache than the Aurora, a massive complex with parks, flower beds, a private beach, tennis courts, a full medical complex and accommodations for 500 people. The resort offered all the facilities any Russian, Eastern European or Central Asian could possibly desire: Swedish massage, Scottish massage, water massage, urological prostate massage, swimming pools, colonic irrigation, saunas, thermal baths, steam cabinets, acupuncture, and eight varieties of mud baths.

Kazakhs wearing heavy gold jewelry crossed the border in their BMWs, and Russians who'd come down for the full two-week medical treatment paraded around the grounds in track suits, bopping to the latest Moscow pop echoing out of the snack bar. Everyone in Kyrgyzstan was waiting for westerners to discover their paradise.

After friends told us about their weekend at Kyrgyzstan's Club Med, we began suspecting that that fantasy was another pipe dream. They'd been pleased to discover, upon arrival, an electric teapot in their room so they could make coffee in the morning or tea before dinner. But when they tried to connect it, they realized that the Japanese plug didn't fit any of the outlets in the room. Having spent years in the former Soviet Union, they didn't waste their breath inquiring whether anyone on the staff had noticed the mismatch or whether the kettle had been meant as pure decoration. They requested a plug adapter. The hotel had none on hand.

For breakfast, they'd been offered a choice of fish or hamburger. When they'd requested bread and jam instead, they were told there was none. "Can we have coffee?" they'd asked their waiter. He'd brought them tea. Mysteriously, guests at other tables were beginning their day with bread, jam and coffee.

The hotel was a behemoth with all the charm of a monument to Stalin. The rooms were Spartan, the service brusque, the decor pure downscale 1950s, and my friends were correct, the food an atrocious attempt at "European cuisine." I ordered eggs for breakfast and was served two raw yolks plunked down in a pool of slightly browned meringue. Lunch was advertised as a chicken galantine; minced mystery meat was served.

Since there weren't more than a hundred other guests on the premises, though, I figured they must have a few extra blankets that would keep me from solidifying in my bed.

"Excuse me, can I arrange to get some extra blankets?" I asked the guard at the door. A rifle at his side, he was wearing pieces of his old Soviet Army uniform.

"*Shto?*" he responded, his English clearly nonexistent. The only staff member we'd met who spoke English was our waiter, who could understand the words *eggs, chicken, meat, no menu,* and *no coffee.* Even the list of spa offerings was Russian-only in Kyrgyzstan's Year of Tourism.

I thumbed through my dictionary and tried to get my tongue around *sherstyanoe odeyalo,* "wool blanket."

He grimaced briefly, trying to find Russian encased in the American accent. When his face finally lit up with a marked, if disinterested, "*Dah,*" I produced my room key so he would know where my relief should be delivered. Instead, he shook his head and muttered what I took to mean, "There's no maid service this late at night." Then he began drawing me a map of the bowels of the hotel.

As I headed toward the elevator, I spied a figure wandering aimlessly, gazing at the colossal chandelier, at the Soviet-era murals on the walls, appreciating—admiring—the oversize grandeur I found so hopelessly tacky. It was Azamat, in reverie. It was his first time as a guest in a hotel, and as he said, "This is Aurora, not any hotel."

One foot planted firmly in the distant past, another just as securely in more recently bygone days, Azamat nonetheless yearned for an idealized Western future, a game of cultural Twister at once poignant and excruciating.

My heart ached as I watched, unnoticed, the wonder playing across his face, the promise not just of comfort but of sights unseen and the excitement of possibility. Like young people from so many parts of the globe, Azamat knew nothing about the unwritten rules of that game: that you can't begin to touch your glittering fantasy without shifting your weight.

18

Iraq and a Hard Place

Don't sleep, you'll freeze.
—KAZAKH PROVERB

Even at 10:30 P.M. on a weekday night, every table at the Qamar al-Zaman restaurant was jammed with customers and groaning with food. Families were polishing off heaping platters of lamb kebabs, barbequed ribs, roast chicken and *guzi laham*, a sumptuous mixture of lamb, almonds and raisins baked in a pastry shell. Businessmen dipped steaming rounds of naan bread into dishes of mezze, small servings of hummus, stuffed olives, tomatoes and cucumber salad, fava beans in tomato sauce, baba ganoush and tabouli. And couples tarried languidly over coffee and tea.

Lines of cars circled Abu Nuwas Street for parking spots by one of a dozen downtown ice cream shops, obese husbands and wives—in a country where half the adults were considered over-weight—dragging along perfectly dressed children for a butterfat break from the stifling 101-degree heat. Late-model BMWs honked incessantly at decrepit Volkswagens blocked by brand-new double-decker buses.

Dozens of trucks wound through the city, delivering rebar and cement for scores of new homes—McMansions with Arabian Nights motifs. At the Shorja market, vendors lined up with crates of Orange Fanta, shampoo, hair gel, peroxide, pirated CDs, DVDs and cassettes of the latest American movies. In the Karata neighborhood, the well-to-do sampled pasta at upscale Italian restaurants and shopped for the latest fashions imported from Milan and Paris. At pool halls in meaner neighborhoods, teenagers ducked and bobbed at Playstation centers, trying to become super ninja heroes or football stars.

Sparkling Mercedes and Peugeots still exuding the smell of new leather pulled up by the front door of the Al Rashid Hotel, their well-to-do drivers and passengers, dressed in Italian suits, smiling as they stepped across the welcome mat, a tiled mosaic of George Bush Sr. captioned "Bush is Criminal" in both English and Arabic—although you would have thought the thrill would have worn off after more than a decade, or at least since his son became the real enemy. Inside, they swished past the ubiquitous portraits of His Excellency, through a lobby with an orange-and-brown 1970s decor, to the bar, or what used to be the bar, for a hookah and some Arabic pop music. In the old days, they had served alcoholic drinks, just as in the old days the hotel disco had always set aside a table for Udai Hussein, Saddam's thuggish elder son, who had once beaten his father's valet to death at a reception for Suzanne Mubarak, the wife of the president of Egypt. But although you could still buy a bottle of 80-proof vodka at any of a score of liquor stores, by decree of His Excellency, alcohol had not been served in public since 1994 and the discos had all been closed.

I'm not entirely sure what I expected to see in Baghdad, but after years of news reports about children malnourished and sickened by UN-imposed sanctions, it certainly wasn't portly businessmen and their meticulously coiffed wives lazing in cafés discussing the latest Julia Roberts movie.

Iraq had not been on the itinerary that we'd laid out for ourselves more than a year earlier, and who in their right mind would go to Iraq in May, when the temperature regularly hovered above 100 degrees, soaring to 110 or higher under the baking midday sun, and where the air-conditioning, where it existed, rarely worked very well? But the day after we arrived home from Iran, President Bush had declared the Axis of Evil, and as residents of the old Evil Empire, we found the prospect of traversing the final two countries on the map of malevolence irresistible.

North Korea was eluding us because Kim Jong-Il's government refused visas to all Americans. "Buy a Kyrgyz passport?" my students had proposed. I was titillated at the prospect of purchasing pseudo-citizenship although worried how I'd pass for Kyrgyz when I didn't even speak Russian. "It's no problem," my friend Vitale reassured me, promising to check how big a bribe the passport clerks would demand. "That's what I love about my country: with money, anything is possible."

Arranging a trip to Iraq had proven easier, although it had been a bundle of contradictions. Most Americans assume that freedom to travel is a thread woven into the fabric of our constitutional rights, but that supposition is both myth and conceit. The U.S. government has banned the use of American passports for travel to Libya, visits to Cuba require special permission from the Department of the Treasury, and Iraq is strictly off-limits, violators subject to fines and/or imprisonment.

Theoretically, journalists were exempt from that provision, but the State Department had adopted a narrow definition of *journalist* that excluded anyone, like me, who was self-employed. When my attorney called the U.S. government official in charge of Iraq travel matters, she told him, explicitly, that should immigration agents see an Iraq stamp on my passport, it would be seized and I would be prosecuted.

If the State Department had deemed me a journalist, I would have been caught in a different trap because the Iraqis weren't granting many visas to American journalists and when they were, the reporters were kept on a tight leash. The Iraqi government, however, was perfectly willing to grant us tourist visas so long as we were part of a group of at least five who traveled with a guide. Tour guides, I assumed, would be both more voluble and easier to evade than the official minders assigned to eavesdrop on media types. When I learned that a British travel agent was putting together the first organized tour to Iraq for Western tourists in four years—and that the Iraqis were only too happy to help me circumvent American law by stamping our visas onto a plain piece of paper rather than inside our passports—we began dreaming of Baghdad.

Our only real fear was that the United States would invade while we were walking the ruins of Babylon or traipsing through Abraham's old homestead at Ur, so we turned for guidance to our friend Steve, a member of the U.S. Special Forces. Steve is not his real name, a necessary precaution for a man who has lived a life built around "plausible deniability." One of those men whose eyes constantly survey the crowds around them, Steve was in Kyrgyzstan conducting counterterrorism training with his Kyrgyz counterparts. After seventeen years holding her breath as her husband flew off to rescue ambassadors from riots in West Africa and "keep the peace" in Bosnia, Steve's wife wanted him to retire before his luck ran out. "This is the wrong time," he explained. "This is when everything I've learned to do can make a real difference."

One night over dinner, we brought up our Iraq fantasy. "So, do you think that we're safe from a U.S. invasion if we get out of Iraq by June 1?" I asked him.

"Off the record?" he asked. "I don't know. My gut feeling? You're safe through December."

With that scant reassurance, we signed up for the tour, flew to Amman, Jordan, and joined a small group—we were three Americans and five Brits—for a fifteen-hour bus ride across a desert so empty that it made New Mexico look lush and overpopulated. At the border checkpoint, a paramilitary strip mall of offices filled with sagging furniture, Iraqi officials politely gathered our papers and pointed us toward a café filled with scores of truck drivers sipping tea while their vehicles were searched. Our luggage underwent the same treatment, each piece of paper and underwear, every shoe and bottle of conditioner, examined, squeezed and considered. While we had our blood drawn and tested for HIV, customs inspectors rifled through our bus—through the engine compartment and the back corners of the overhead bins. Finally, we were instructed to produce our electronic equipment. Cellphones and video cameras were confiscated, with a promise of their return upon our departure. Still cameras were "registered" in longhand by a clerk whose literacy turned the process into a two-hour marathon.

It was a warm welcome compared with what would have greeted us at Baghdad's Saddam International Airport, where blood-red DOWN WITH AMERICA slogans decorate the gangways.

We hadn't driven twelve miles into Iraq before we realized that the sanctions, highly touted both by Iraq and the United States, had more holes in them than a spaghetti strainer. A second road paralleled the highway leading across the barren wilderness to Baghdad, a special freeway for the steady stream of tankers, trucks and refitted buses that carried oil in violation of the United Nations sanction against Iraqi sales of crude. At what passed for a truck stop—a modern restaurant oasis—a kiosk offered Cokes imported from Syria, Orange Crush from Iran, Kents from Jordan and Marlboro Lights marked with no point of origin. "From Israel," our guide whispered, "but if the packages said that, no one would buy them."

The highway to Jordan wasn't Iraq's only route for evading sanctions. The Iraqis were trucking oil to Turkey, pumping it by pipeline to Syria and smuggling it through the Shatt-al-Arab waterway into the Gulf.

Iraq had mounted a remarkably successful international public relations campaign that had limned a stark portrait of the human cost of sanctions: children turned listless by hunger, the water supply polluted because basic chemicals like chlorine and iodine were on the list of prohibited purchases, the sick suffering in hospitals without medicine. But between the smuggling and the rising price of crude, which generated Iraq an extra $1 million a day for every dollar increase per barrel, the Iraqi economy was expanding at the rate of 15 percent annually. The government was spending only two-thirds of the revenue the UN oil-for-food sanctions program had pegged for sanitation and just half of its allocation for health care. And it had just declared a one-month oil embargo to protest the aggression of "the Zionist entity," as the *Iraq Daily* had explained.

Unlike Fidel Castro, who makes it a habit of showing up unannounced at schools and in remote villages, Saddam was so reclusive that few Iraqis ever saw him in the flesh. His Excellency was nonetheless ubiquitous, Saddam billboards the dominant iconography of Iraqi life, appearing like advertisements for the Gap or Banana Republic. But Saddam was no Turkmenbashi, who displayed himself in but a single outfit and a single set of jewelry, or even Lenin, whose wardrobe varied but whose demeanor never wavered. Saddam wore more faces than Imelda Marcos wore shoes: stern-faced Saddam in the uniform of a field marshal; rakish Saddam in a leather jacket and dark shades; impartial Saddam in a judge's robes balancing the scales of justice; fatherly Saddam at a hospital bedside; wise Saddam in a mullah's robes; casual Saddam in a sweater vest talking on the telephone; relaxed Saddam in a straw boater; Saddam the hunter, his rifle aloft; Saddam the planter, wielding a hoe; and diplomatic Saddam in gold-trimmed robes, hugging his faceless Arab brethren.

Baghdad's triumphal arches, twenty-four-ton swords forged from the melted and recast guns of dead Iraqi soldiers, were held aloft by two fearsome fists, copied detail for detail from Saddam's own hands. At the Saddam Hussein—or Triumphant Leader—Museum, you could admire a sampling of His Excellency's favorite weaponry, from pistols and rifles to machine guns and shotguns; his dictator drag, uniforms in a dozen colors and permutations; even the scores of cards, letters and gifts he'd received from his pals, including Fidel, who'd mysteriously sent six dress shirts. You could even gape in awe at the replica of a 605-page Holy Koran written in 24 liters of Saddam's own blood, the original in the Umm al-Ma'arik—the Mother of All Battles—Mosque.

A statue of Saddam welcomed you to Saddam Tower, a 600-foot telecommunications building. This was Saddam the fearless, twice as big as life, in an open-collared shirt, gloating over the shards of American cruise missiles scattered at his feet. From the revolving rooftop restaurant, you looked down on Saddam's Baghdad Palace, which made the White House look like an outhouse in rural Alabama. But, caution, no photographs, lest those dastardly Americans send spies to photograph a complex easily captured by cameras in space. Far in the distance, you could catch a glimpse of the skeleton and limestone arches of the city's newest holy site, still under construction, the Saddam Mosque, planned as the third largest mosque in the world, after Mecca and Medina, demanding more rebar than the Empire State Building, our guides excitedly informed us.

A new play based on Saddam's first novel, *Zabibah and the King*, had just opened at Iraq's National Theater, directed by Sami Abdul Hamid, who trained at the Royal Academy in London and held a doctorate from the University of Oregon. The musical drama was an allegory disguised as a romance, the story of a noble king who falls in love with a beautiful peasant woman—a stand-in for the Iraqi people—who is raped by her brutish husband, coincidentally on the very date the Gulf War began. The king declares war to avenge her honor in a scene choreographed with a macabre dance performed by soldiers in gas masks. But Saddam offers no "happily ever after." Zabibah dies, but inspired by her love, the king gives more political freedom to his people, or at least he tries. Sadly, their representatives sink into petty squabbling. As the curtain closes, the king, too, succumbs, leaving the people uncertain how the country can survive the loss of their beloved father figure.

Each cultist of his own personality has a historical role model. Turkmenbashi was a wannabe Lenin, and Karimov did a pretty good imitation of Timur. It didn't take long to realize that Saddam reached farther back in history for his inspiration. His genealogists were ordered to construct a plausible family tree linking him to Fatima, the daughter of the prophet Muhammad. On the huge replicas of his head that surrounded his Baghdad palace, Saddam wore the helmet of Saladin, the great Islamic victor in the Crusades, also a native of Tikrit, Saddam's hometown. And at the ruins of Hatra, the ancient Assyrian fortress city that dominated the caravan trade in northwest Iraq from the fourth century B.C.E. to A.D. 300, the major temple, dedicated to the sun god, had been reconstructed, each brick carefully stamped, IN THE ERA OF SADDAM HUSSEIN.

But it was at Babylon that Saddam claimed his true birthright by building a palace on the crest of an artificial hill overlooking the site of the first city-state on the planet, where the first human laws were codified. Below it—beneath it, as if part of his grounds—sat the palace of Nebuchadnezzar II, which Saddam had ordered rebuilt. Over the old bricks, marked with Nebuchadnezzar's name, a new set had been laid, declaring, IN THE ERA OF SADDAM HUSSEIN, PROTECTOR OF IRAQ, WHO REBUILT THE ROYAL PALACE.

Saddam gloried in Babylon each September when he hosted the Babylon International Festival, its slogan "From Nebuchadnezzar to Saddam Hussein, Babylon rises anew on the path of construction and glorious jihad." The ancient city, the festival's organizers advertised, represents "the great cultural achievements of great historical figures, such as Hammurabi, and Nebuchadnezzar, whose legacies transcended time and history to fuse with the great strides, and magnificent splendor of H. E. president Saddam Hussein. . . . Like the mythical phoenix, it overcomes embargoes to resume its journey of defiance and accomplishment."

Before their expulsion in 1998, United Nations weapons inspectors had driven to Babylon to tour Saddam's palace but were denied entry. Instead, they spent the day touring Nebuchadnezzar's.

Iraqis didn't know what to make of our small band of travelers.

"Are you on a solidarity mission?" asked doormen and hotel clerks accustomed to Europeans and Americans bringing books and medical supplies to Baghdad or showing up to parade their opposition to sanctions. The week we arrived, a delegation from the U.S. group Physicians for Social Responsibility appeared on television swooning at the feet of Tariq Aziz, the foreign minister. And a Veterans for Peace brigade came bearing STOP THE OCCUPATION—PEACE AND JUSTICE FOR PALESTINIANS T-shirts and $40,000 of the $60,000 needed to rebuild water treatment plants in the cities of Falooja and Baaqooba.

"No, we're tourists," we responded.

"Tourists?"

The closest thing to tourists Iraqis had seen in almost a decade were Muslims from Iran, Bahrain, India and Iraq on pilgrimages to holy Shiite religious sites. One look at my uncovered head was enough to tell anyone that I wasn't in Iraq to pray at the tomb of a prophet.

The Al-Rashid Hotel, Baghdad's finest, was still packed. But the men—always men—lazing by the swimming pools or eating $30 breakfasts were guests of the government, European businessmen and journalists, not leisure travelers. They were utterly indifferent to the stunning selection of Saddam Hussein watches—gold, silver, digital or winding—Saddam Hussein key chains and Saddam Hussein carpets on sale at the upscale boutiques in the hotel's shopping arcade. We, of course, couldn't resist expanding our dictator collection.

The solidarity crowd gathered at the Palestine, a lower-scale five-star hotel. But since they were busy offering succor, the advertised miniature golf course had been abandoned and the old bowling alley was closed. Even the outdoor tourist restaurants along the banks of the Tigris, famous for their *masgouf*, a flat fish split open and roasted on stakes, were empty of all but a handful of locals.

Each morning, the owners of shops and stalls in Baghdad's tourist bazaar pulled up their steel shutters and turned on their lights. And at the end of the day, having made no sales whatsoever, they reversed the process and sighed the bottomlessly fatalistic Arab expression, "*Inshallah* [God willing], tomorrow will be better."

Accustomed to buses draped with banners reading END THE RACIST SANCTIONS NOW! and LET IRAQ LIVE in English and Arabic, the somber guide at the al-Amiriya air raid shelter was taken aback by our small cadre when we pulled up in a new minivan and trooped into the charred room without the obligatory wreath of condolence.

Our guides had done their best to dissuade us from visiting the bomb shelter, a macabre shrine of broken concrete and twisted steel that was an obligatory stop on any solidarity tour of Iraq, and we didn't quite understand their hesitation. All we knew was the story we'd read in our guidebook and heard from antisanctions activists trembling with fury at the inhumanity of the United States: On the night of February 13, 1991, more than 400 women and children who'd taken shelter there from American attacks had been incinerated when a two-ton bomb pierced five layers of concrete above them, melting the structural rebar into a tangle and turning their refuge into an inferno.

It was obvious that something horrendous had happened inside the windowless blackened room, opened to the blazing desert sun by a gaping hole of twisted metal and shattered concrete. But the Iraqis weren't content to let the destruction speak for itself, or to answer obvious questions like: Why was the shelter at ground level rather than buried? Why were earlier groups told that 1,200 people had died? Why were there only thirty-four shelters in a city of 4 million?

"See here," said our guide, pointing to hundreds of blackened slashes on the ceiling. "Look what happened to our children. Those are the charred hand prints of the babies who were sleeping in the top bunks. This was a crime against humanity."

The marks on the ceiling were curiously identical, as if every child in the room had suffered the same ghastly fate while poised in the exact same position. Why were there hundreds and hundreds of them? And why would mothers have put their kids in the top bunks? The grief of the guide, who told us that she'd lost her own children in the shelter, was palpable. She'd even berated me for chewing gum. It was impossible, then, to voice any skepticism.

As we wandered the blackened room, glancing at the photographs of the victims, the wreaths and letters of support and condolence left by activists from Malaysia and France, Spain and the United States, I recalled the accounts of the devastation written by members of various peace delegations. The first bomb had cut off the air supply and sealed the forty-ton exit doors, they'd said, and the women and children who hadn't been killed by the blast had been boiled alive when a second bomb burst the shelter's water tanks, releasing steam at 2,000 degrees Fahrenheit.

"Where were the water tanks?" I asked, realizing that I hadn't seen any likely hunks of metal.

"No, that story wasn't true." A note of hostility had crept into her voice. The survivors, she contended, had been burned alive when the U.S. pilots used the hole created by the missile to douse survivors with napalm—all part of an intentional plan of genocide.

Even the less-than-skeptical Brits in our group blanched at that assertion. "How do you know it wasn't a tragic accident?" asked one retired physician. The Pentagon had claimed just that, that they had selected the site believing it to be a command-and-control center.

"We know," asserted the guide, confidently. "His Excellency told us that the Americans targeted all of the air raid shelters, but the others were saved because the world was so horrified by what happened here that Bush was forced to end the war."

The sun was just setting over the Babylon River when our bus pulled up to the Babel Hotel after a long, hot afternoon traipsing the reconstructed ruins at Babylon.

I'm not a religious Jew, but as we followed roads that wound back and forth across the Euphrates River, history stirred inside me. While Dennis checked in to the hotel, then, I performed what surely must have been a ritual for Jews visiting Iraq for centuries: I sat by the waters of Babylon—in my modern rite humming Bob Marley's song, banned in Iraq because of the word *Zion*—and wept.

That night after dinner, our guide, who I'm calling Omar to shroud his identity, brought his tea to our table. "You are Christian?" he asked. A scion of the Iraqi upper middle class, Omar had studied in Pakistan and India, skied in Gstaad and sampled the finest cuisine of London. In the old days, Dubai had been his paradise, but no longer. "They don't treat Iraqis with respect there anymore," he commented, so certain that I would sympathize with his plight that he didn't even try to disguise the fact that he considered serving as a guide to be beneath him. "They make you ashamed."

Omar had homed in on me as soon as he joined our small band, I assumed because I was the only woman in the group. Obsessed with sex, he seemed incapable of resisting the opportunity to discuss the unmentionable with an actual female.

"Veiled women are the best," he confided. "They pretend they're pure, but that's a cover for the truth that they are the most passionate of all women." And he bragged about the sexy clothing he bought his wife to wear at home: "I would never let anyone else see her dress that way, of course, but I have the right to that pleasure."

Earlier that day, he'd asked, "It is true that American women are not required to be virgins when they marry?" We'd just watched a wedding party enter our hotel, a procession of family members who all followed the terrified-looking bride and the beaming, wide-eyed groom upstairs in the elevator.

"Why don't they leave them alone?" I'd inquired.

"Ah, they need to see the sheet, you know, to make sure that the bride is pure," Omar explained.

"What about the groom?" I'd decided it was time to poke a bit at his preconceptions.

"But men aren't virgins or not virgins. Men aren't supposed to be pure."

With Dennis, Omar had discussed the pornography he'd bought from Palestinian smugglers, salivating over the variety of sexual possibilities. And, just like the men Dennis had met in Afghanistan, he whispered and giggled repeatedly about man-on-man sex with the fascination and horror of an eleven-year-old boy. "Do you know about homosexuality?" he asked one afternoon, as if only Iraqis were savvy to this exotic form of sexual behavior. "Men who can't afford prostitutes meet each other for sex at saunas and Turkish baths. Do they do that in America?"

Dennis, too, found Omar's pretensions to hipness an irresistible lure. "In America, many think men who hold hands with each other and kiss each other are homosexuals," he egged him on, recalling having seen Omar do both. "And remember the Turkish bath at the hotel, when the men were soaping each other down? In the West, many people would assume they were also homosexuals. And what about that guy who told Elli that her husband was handsome? Do you think he was a homosexual?"

Chilled by the implications, Omar's face turned a deep shade of purple as he sputtered out a firm, "Of course not!"

Before dinner in Babylon, I'd quizzed Omar about the position of women in Iraqi public life, in the military, in politics and business.

"Women are too emotional to be leaders or judges," he opined in a fatherly voice, as if instructing a child.

"What about Madeleine Albright?"

"She's not a real woman."

"Benazir Bhutto?" I tried the former prime minister of Pakistan.

"She's a lesbian."

There was no point in continuing, so I buried my face in a guidebook. That's when Omar asked me whether I was Christian.

Jews traveling in Muslim countries don't have the luxury of being casual about their religion and ethnicity. When we fill out visa applications and fall into casual conversation, we have to decide whether to lie or to run the risk of provocation. Having more talent for the latter than the former, I replied to Omar's original query bluntly, "No, I'm not Christian, I'm Jewish."

Omar didn't blink an eye. "There used to be many Jews in Iraq," he said in the same lecturing tone he used to instruct non-Muslims about the historical division between the Sunni and the Shiites.

In fact, the thirty-eight Jews left in Baghdad were the survivors of a Jewish community that was more than 2,500 years old, dating from the exile of the Jews to Babylon after the Chaldean king Nebuchadnezzar laid waste to Jerusalem and destroyed King Solomon's Temple. Jewish history is inextricably bound up with Babylon, the birthplace of Abraham, Sarah and Rachel, the site of the tombs of Jonah, Ezekiel and Ezra. More than half a millennium before the birth of Christianity, and more than 1,200 years before the rise of Islam, the Jews of Babylon were shaping the final version of the Torah, writing the book of Job, and compiling the Talmud, the canon of Jewish law and its interpretations.

For the most part, Jews had thrived in the Muslim country until a Mufti-inspired pro-Nazi coup in 1941 ignited a series of pogroms that left 180 dead and 1,000 wounded. Then, in 1950, Iraq's Jews were "encouraged" to leave and forfeit their citizenship. But after more than 100,000 did, in a clandestine Israeli airlift dubbed "Operation Ezra and Nehemiah," the Iraqi government slammed the door shut on the remaining 25,000 and required them to carry yellow identity cards. Following the Six Day War in 1967, Jewish property in Iraq was expropriated, Jews were dismissed from posts in the government, their telephones were disconnected and their bank accounts frozen.

Anti-Zionist sentiment was maintained at a fever pitch, a strange counterpoint to the popular distaste for the Palestinians themselves—widely considered to be "trash Arabs," as many locals put it. Sandwiched in between stories cribbed from the BBC about Madonna's latest endeavors and reviews of new American films like *Undercover Brother*, the *Iraq Daily*, the country's only English-language paper, ran headlines like, ZIONISTS DRINK HUMAN BLOOD.

The anti-Zionism was predictable; the overt anti-Semitism was more chilling. From their pulpits, Iraqi imams preached overt anti-Semitism, as in "O God, have mercy on the martyrs. O God, destroy the Jews." In 2000, the Iraq government labeled the Jews "descendents of monkeys and pigs, and worshippers of the infidel tyrant," and the newspaper *Babil*, owned by Saddam's eldest son, Uday, ran a piece about the Jews of his own country. "The Jew by nature loves material things," it read. "He differs from all other peoples living in the region by his materialistic attitude. . . . Ever since the Jews arrived in Iraq, their prime concern was to gain control of Iraqi commerce."

Ordinary Iraqis, then—even educated Iraqis—were steeped in politically created ignorance about all matters Jewish. During our travels, I discovered in conversations with Iraqis that Jewish women are required to lose their virginity before marriage and that Jewish permit stealing so long as the thief doesn't get caught. The Wailing Wall, I learned, was the remains of a nineteenth-century building that had no historical significance, and Israeli archaeologists were digging under it not to unearth valuable historical information but to collapse Al-Aqsa Mosque. Oh, and the two blue stripes on the Israeli flag—designed in the late nineteenth century to suggest the blue stripes on the tallith, the traditional Jewish prayer shawl—really represented the Nile and the Euphrates Rivers, the expanded borders Greater Israel was intent on establishing.

The Iraqis weren't whipped up to anti-Semitism by religious fervor. In urban areas, they were the most secular of Arab Muslims. Mosques were sparsely attended, especially in comparison to the crush in Pakistan, for example, and we saw no one stop working and pull out a prayer rug for the five-times daily *salaat*, as we had even in cities in West Africa. In villages in the South, men commonly had two or three wives. But few women outside of those settlements wore veils, as they would have on the other side of the border in Iran.

Saddam, a Sunni Muslim, had long been a secular ruler. During Shiite antigovernment rebellions in 1991, he'd directed his military to bomb a mosque in Kerbala, where rebel Shiites had taken shelter. But in a bid for legitimacy in the Arab world, Saddam had recently gotten religion. He'd banned alcohol from public places, sponsored talent competitions for the best chanters of the Koran and stepped up an old "Back to the Faith" campaign. Transferring his obsession with building extravagant presidential palaces to an obsession with building extravagant mosques, he'd set aside $4 billion for the Saddam Mosque alone.

Our guide Omar was one of dozens of upper-middle-class men who bemoaned this new spate of religiosity. Their Baghdad, the Baghdad of the 1970s and '80s, had been filled with risqué nightclubs where women from Thailand and the Philippines bared most of their flesh. Wealthy Arabs from Kuwait and the Emirates had flocked to casinos in the South and to the discos of Baghdad.

"Now my wife can't even find short-sleeved dresses or blouses to buy," Omar complained. "Women feel that they have to cover their arms and their beautiful hair.

"Even though His Excellency is a Sunni, the Shiites are winning."

In Bishkek, Tashkent and Ashgabat, the questions came at me like machine-gun fire: Why does the United States intervene in other countries? What do Americans think of Central Asia? Why did Americans cheat at the Olympics? In Afghanistan, I couldn't walk two blocks without being pulled into a conversation about the United States government or Bush's foreign policy. And in Iran, every encounter turned into a dialogue about why the United States hated Iran.

In Iraq, the silence was deafening.

When we checked into the Oberoi Hotel in Mosul—where we were the sole guests—a crowd formed around us, strangers curious about the nationality of the obvious foreigners.

"America! Wonderful! A great country!" they gushed.

Outside the Temple of the Sun in Hatra, the lonely archaeologist who kept up the ruins of a city destroyed 1,800 years ago couldn't contain his shock. "We don't see many Americans," he exclaimed. "We love Americans."

One afternoon, we strolled through the market in Nasiriyah, and scores of men and boys gathered around us, following our every step. Then a friendly barber invited us for tea and insisted on trimming back Dennis's hair.

"Where are you from?" he asked, as half the marketgoers and shopkeepers listened at the door.

"What do you do for a living?" one yelled.

"How many children do you have?" asked another.

"Do you like Iraq?"

As we watched the sun set over the Tigris one scorching evening, a group of women—the eldest shrouded in black, the youngest sporting light skirts and blouses—motioned me to join them, offering me a chunk of their bread, a slice of their melon. I noticed one of them flinch when I uttered the word *America*. But they seemed more concerned with the cut of my dress and the quality of their hospitality than in discussing our nations' hostilities.

Only once—when we were paying for souvenirs at the gift shop at the ruins of Nebuchadnezzar's palace—did an Iraqi mention the not-so-cold war between America and Iraq. Counting up the 32,000 dinar that added up to $10, the archaeologist at the site, who doubled as a tour guide, complained, "It's your country's fault that our money is like this."

Other than that passing comment, you would never have known that the U.S. government had bombed half the bridges in the country, that U.S. warplanes were enforcing the no-fly zones, or that Bush had declared Saddam dictator *non grata*.

Unwilling to leave Iraq without gauging popular sentiments— I'm a journalist, after all, and journalists rarely content themselves with ruins—I tried backing into politics.

"It's so sad that our two countries don't get along." Or: "The Iraqi people and the American people should know each other better."

Nothing.

I stepped up my inquiries.

"What impact have the sanctions had on your life?"

That, at least, provoked some mild response. My favorite: "Now, thieves and smugglers rise to the top instead of trained professionals," said Omar. "My brother-in-law, who has a sixth-grade education, is a multimillionaire. I have an MS and speak four languages and have to work as a tour guide."

Finally, I became blunt, asking every person I encountered, "Are you aware that President Bush has declared Iraq to be part of an Axis of Evil and that he's talking about invading in order to overthrow your president?"

"Yes," they responded, without much emotion.

"This is normal for us, Iraqis have been at war for the last twenty-two years," several bus drivers explained.

"We know it will happen, but what can we do about it?" others replied with the total resignation Americans reserve for the annual increase in gas prices.

"It must happen," an Assyrian from Babylon, a Christian desperate to leave the country, told me. "And it is America's fault for not finishing the job in 1991."

Although opinions varied on dozens of political topics, from religion to how much Iraq owed the Palestinians, one belief united everyone I met, Shiite or Sunni, urban or rural, Christian or Muslim: everything was the fault of the United States.

- Saddam Hussein invaded and annexed Kuwait, but that invasion was the fault of the United States, either because the United States was conspiring with Kuwait to lower oil prices, which was destroying the Iraqi economy, or because the U.S. ambassador, April Glaspie, gave Saddam the green light to invade.

- At a cost of more than $2 billion, Saddam Hussein built dozens of "palaces"—the joke in Iraq being that he was building one for each of Iraq's 24 million citizens: the Green Palace at Lake Tharthar that covers two and a half

square miles; Qasr-Shatt-al-Arab on the waterway between Iran and Iraq, bigger than the palace at Versailles; the
presidential compound at Mosul, which includes three
lakes and man-made waterfalls; Radwaniyah, west of
Baghdad, home to four presidential residences and 225
VIP villas. But Iraqi children drank polluted water because
the United States was engaging in economic warfare
against Iraq.

- Saddam was sending $25,000 to the families of all the
  Palestinian suicide bombers, but 4 to 6 percent of Iraqi
  adults were malnourished and 20 percent of its children
  underweight because the United States would not allow
  the country to earn enough money to buy them food.

- Saddam declared a thirty-day moratorium on Iraqi oil sales
  to protest Israeli aggression, but because the United States
  supported Israel, it was America's fault that Iraqi schools
  had no money for books.

- There were cracks on the Northwest Palace of King
  Ashur-nasir-pal II at Nimrud. That, too, was America's
  fault, although the palace was built about 850 B.C.E.

To be fair, some things were the fault of Iran, like the deaths of
375,000 Iraqis in the Iran-Iraq War. But despite the fact that Iraq
invaded Iran and that, by the end of the war, the United States was
actually helping Saddam Hussein, even the blame for that carnage
was laid at the feet of America, although no one could clarify that
line of thinking.

At first, I assumed that the almost ritual laying of blame was
Iraqi caution. Saddam, after all, wasn't notorious for his patience
with dissent. All those soldiers, police and guards at intersections or
at checkpoints on even the most remote rural roads weren't waiting
around for an invasion. Authoritarianism was such a fact of Iraqi life
that no one found it strange that the only available road maps were
those foreigners downloaded from the CIA website.

But while Iraqis trod the cautious dance of the fearful in conversation, they seemed more ignorant than afraid. After all, they

were incredibly cut off, even by the standards of the Taliban's Afghanistan, where magazines, people and ideas had always flowed across the border from Pakistan. In Iraq, however, the import of foreign publications was forbidden, and the few well-heeled Iraqis permitted to travel abroad left their relatives behind as guarantees of their good behavior. Possessing a satellite dish or satellite phone was a crime punishable by imprisonment. Telephone lines were tapped and on-line access was limited to government offices, a handful of private businesses and special Internet centers, where all computers faced away from the wall so that officials from the government's Internet department could peer over your shoulder to make sure you didn't use an unapproved e-mail account or access a forbidden website.

Iraqis learned, then, only what the government wanted them to know—and that wasn't much. Even local journalists couldn't stand the pressure. In 2001 alone, fifty journalists fled a country where typewriters had to be registered with the government. Night after night, then, Iraqis were fed a steady diet of news about "the American administration of evil" and about the attacks by "the Zionist entity" against the martyred Palestinians, coordinated, of course, by Washington. Even those open in their condemnation of Saddam Hussein were foundering in a sea of dis-, mis- and non-information. "Why did so many Jews who worked at the World Trade Center stay home from work on September 11 if the Jews weren't responsible for the attack?" one young dissident asked me. When I explained that his information was surely incorrect, he shook his head vigorously. "I have heard that reported many times."

The story the Iraqis had been taught for almost a quarter of a century was that their great nation had been assaulted time and again, by the Greeks and the Assyrians, Armenians, Philistines, Persians, Mongols, Turks and British. The Americans were just the latest foolhardy imperialists with their greedy eyes fixed on the rich land between the lower Tigris and the Euphrates, and they would meet the same fate as their predecessors.

"Iraq isn't Afghanistan," an earnest young Iraqi Christian—a member of a community 350,000 strong—warned me. Since he couldn't have despised Saddam more if His Excellency were either

George Bush or Ariel Sharon, he was bemoaning that reality. "It's not just that our military is well-trained and experienced. The Taliban was in power there only five years. Saddam has controlled Iraq for my entire life."

I only realized how eerie it was to be a tourist in a country my nation was about to bomb when I found myself drinking tea at the top of Saddam's tower in Baghdad, thinking, "This will be one of the first structures to go," or wondering how much longer the bridge I was crossing over the Euphrates would stand. If the nice young barber in the market in Kerbala who cut Dennis's hair would survive the attack. Whether the woman who'd smiled so sweetly when she sold me a Snicker's bar to sate my hunger during a long wait for dinner would spit in my face if I came back a year later. What would happen to the Al-Rashid Hotel once war began.

My disquiet was most acute in the ancient city of Basra on the Shatt-al-Arab, where the Tigris and Euphrates meet before running into the Gulf. Just six miles from the border with Iran and thirty-five miles from Kuwait, the old Venice of the East where Sinbad the Sailor began his adventures had been the focal point of the Arab sea trade for more than a millennium, and was Iraq's only outlet to the sea. Basrans have a saying, "If there was a war between France and Germany, Basra would be bombed," and over the past century alone, it has endured occupation by the Ottomans and the British, borne the brunt of the eight-year war with Iran, been bombed and strafed from the air, land and sea during the Gulf War and was sinking under the grip of international economic sanctions.

Unlike Baghdad, where nearly every scar left by U.S. bombers had been erased, Basra, once a lively tourist destination thanks to the best casinos in the Gulf, languished, in punishment, all assumed, for the popular uprising that broke out there at the end of the Gulf War. The power grid, decimated by U.S. air strikes in 1991, had not been entirely rebuilt. The water wasn't drinkable, the hospitals were overwhelmed by the rise in congenital birth defects, and the streets were muddy, open sewers.

Along the waterfront, young men with nothing to do but

bemoan their fate lazed by the river while water buffalo plodded through the reeds for relief from the heat and jet-skiers wove in the wake of old flatboats. Behind them towered the Corniche of Mar- tyrs, 101 statues, each a three-times-life-size likeness of a martyr killed in the war with Iran. All but one pointed threateningly, or accusingly, toward Iran. Inexplicably, the final figure, General Adnan Khairallah, one of Iraq's most popular generals, who died "tragi- cally" in a helicopter crash not long after the armistice, seemed to aim his accusation in the opposite direction, toward Baghdad.

The desolate lobby of the Basra Sheraton—which clung to that long-severed affiliation by Xeroxing old stationery and guest information cards—echoed with Dionne Warwick's throaty "Do You Know the Way to San Jose." The blankets and carpets in the spacious rooms were all original, the bathroom fixtures corroded and the shower curtains frayed. The air-conditioning barely worked, and the vanity mirror was missing all but two of its twenty-six bulbs.

Downstairs, a bucket caught the rain leaking through the roof into the shopping arcade, where a perfume shop, a confectionary and an antique store were all shut. The only door open belonged to a travel agency. It, too, was closed, but a maritime engineer who'd worked in Russia sat inside, using a friend's computer.

"Would you believe this was once the third best Sheraton in the world?" he asked me, perhaps seeking reassurance that things weren't really as bad as they seemed. They were, and he knew it, so there was no reason to lie.

He offered me tea, and we conversed in a mixture of broken English—his—and broken Russian—mine. "It will not always be like this," he told me, brightening at the prospect. "Basra will come back."

His optimism perplexed me. Bush was demanding that United Nations inspectors be permitted to return to Iraq to rid the rogue nation of its secret caches of chemical, biological and nuclear weapons. Saddam was lambasting the United States for a weapons- of-mass-destruction double standard that prohibited Iraq from developing technology Israel already had. The Iraqi opposition in exile, long cut off from any potential followers, was a national joke.

But the popular longing for normalcy was so keen that it drowned out realistic pessimism.

I flashed back to that conversation on our last afternoon in Iraq, which we spent selecting carpets in the bazaar in Baghdad. I'd spent almost two hours drinking tea with the owner above his shop, an intrinsic part of haggling for a stack of lovely Iraqi kilims unlike any of the Turkish or Iranian kilims I'd seen. Finally, the deal was struck at $300 and a clerk packed up my purchases.

As I departed, the owner called out a final farewell, then added, "Very good price. After, it will be $300 each."

The veneer of modern police states is so seductively normal that it requires daily acts of conscious will to stay attuned to the essence of that normalcy. Men and women go about their business in a familiar way, commuting to the office, trying on shoes in the market, eating grilled fish at busy restaurants without the ring of jackboots on the pavement. When you force yourself to think about it, you suspect that your phone is tapped or that the floor attendants are keeping meticulous note of your comings and goings. But those chilling realities are invisible. And while you catch an occasional glimmer of hesitation—is that fear?—in the voice of a shopkeeper or clerk and feel, more than notice, the suspicious grimace of a soldier, you don't actually see anyone following you or witness locals being hauled off by the gendarmerie.

The face of fascism is rarely that frank, its menace all the more malignant for its pervasive subtlety.

At least, that's what we thought until we tried to leave Iraq.

During our three weeks meandering from Basra to the hills near the Syrian border, we'd been so insulated from despotism, so self-assured in our foreignness, that we'd become casual about its heft. In the back of the bus, we'd imitated His Eminence, joked about what we were convinced would be his imminent demise, and merrily snapped photographs of his palaces, of Scud missile installations and military training camps. The latter was a silly game, of course, albeit a quintessentially American one. We didn't need, or even particularly want, pictures of half-hidden walls, military hardware or distant soldiers on parade grounds. But the minute our minders warned us not to capture those images on film, the urge became irresistible.

If we'd been Iraqis—or Uzbeks or Iranians—trained in the habits of fear, we would have controlled ourselves. But as white New Yorkers, our habits of fear weren't attuned to those wearing uniforms. We had no experience controlling our more frivolous instincts against the awesome might of the State in any matter more compelling than parking.

We didn't give that self-indulgence a second thought until we drove back to the customs station just shy of the Jordanian border and handed over our cameras to the same semiliterate sergeant who'd registered them in painful detail three weeks earlier. He dispensed quickly enough, for an Iraqi official, at least, with our group's pile of Nikons and Canons. But he stopped short when he reached our Hewlett-Packard digital, turning the strange beast sideways and upside down.

"Where does the film go?" he asked casually, with curiosity, nothing more.

I explained the theory of digital cameras, clearly a revelation in that remote outpost, and flipped out the small disk to illustrate how film had been replaced. Then I snapped his photograph and pointed to his visage in the window on the back of the camera. The sergeant scrunched up his face, then ran his fingers through his hair to contain an errant lock. We all laughed when he requested another snapshot to capture his new, improved look.

With no means to examine them, the sergeant had ignored the piles of film canisters inside our companions' camera cases, not even asking what sorts of sights they had photographed. Once he grasped the concept of the digital camera, however, he sensed a new possibility. The next thing we knew, he was insisting on scrutinizing every image we'd captured electronically, one by one.

"Where was this taken?" he asked about a particularly lifeless vista.

"Why did you photograph that billboard?" he inquired about snapshots take from the bus in Baghdad.

"Who is that person?"

"What's that in the background?"

As often as not, we couldn't answer his questions. It wasn't all that easy to distinguish one swath of desert from another, and the billboards often were incidental, backdrops to buildings. Who were the people? Kids, old men and groups of teenagers who'd lined up for their moment of fame. What's that in the background? We hadn't even noticed the backdrop to group portraits.

His displeasure grew palpable, and our blood pressure skyrocketed. We had taken 486 digital photographs in Iraq, and he was only on number 21.

Omar sweated profusely as he translated, suspecting, we thought, that the sergeant was about to find something illegal and haul us all off—guide included—to a military prison.

For three hours, Dennis and I traded places at the side of our interrogator, giving each other frequent, and desperately needed, nicotine breaks. No one else spoke. What was there to say? And there was certainly nothing to do. We couldn't leave, and the sergeant didn't seem the type who'd respond well to wheedling, crying or bribery. We certainly couldn't behave like New Yorkers and barrage him with a recitation of our rights or complaints about the thuggery of officialdom. Call the embassy? Not likely. All we could do was pray, or what passes for prayer among nonbelievers, which suddenly struck me as the only alternative most Iraqis had.

In the middle of Ur, after picture number 87, the camera went dead, the batteries spent from the long session. We held our collective breath.

"Who has more batteries?" demanded the sergeant, making no effort to disguise his annoyance.

We shrugged our shoulders with vigor, and our companions pulled out seven varieties of batteries they knew would not fit inside our HP.

The sergeant dithered for what was probably only a moment but felt like an hour, clearly trying to decide if we had cooked up the malfunction or were feigning our inability to solve the problem. Finally he barked a few words in Arabic, then signed our exit papers and waved us off.

The paranoid side of me, which had flared into full control over the preceding hours, thought I spied a flicker of disappointment in our near-captor's face, a wistful sadness that he'd missed out on the thrill of arresting two Americans who His Excellency could parade as imperialist spies.

The quieter, rational side of my brain noticed the same disappointment, but with a different interpretation of its origin. Stuck out on a border in the middle of nowhere, without so much as a television or a movie theater within ten hours, the sergeant was frustrated that the picture show had been so rudely interrupted.

19

Not Fit to Print

Without a paper you're just a bug;
with a paper, you're a person.
—RUSSIAN PROVERB

"Fascist pornography."

The message delivered by Artyom contained more than those two words, but I was so taken aback by that inflammatory pronouncement that I could barely absorb the rest, the words *banned, yellow journalism* and *Ministry of State Security.*

The first issue of our student newspaper had appeared on International Freedom of the Press Day, after weeks during which my students had handed in sloppy essays thinly disguised as journalism, busted all the deadlines I'd established and dithered endlessly about the name and the design of the masthead.

I hadn't expected that they'd emerge from the process of editing their own newspaper singing "We Shall Overcome" with a Kyrgyz beat, but I'd hoped for the old attitude of *Rolling Stone.* Alas, as I read through the first batch of stories with the help of Regina, my ever-present translator, I realized that I'd gotten the *Utne Reader*. The articles were perfectly acceptable but safe and dull—one about the dangers of smoking, another about an art exhibition, yet another a prosaic profile of me. It was precisely the kind of work their local professors would have applauded.

Except:

Two girls, so quiet that I was sure they'd slept through both semesters, had taken to heart my suggestion that they write about people they knew, and produced an article about skinheads. In a society still gripped by the specter of fascism, even half a century after the defeat of the Nazis, that showed some mettle.

Two other girls—both freshmen, so young enough to retain a shred of optimism, perhaps—had boldly gone where no Kyrgyzstani journalist, never mind student, had gone before. Inspired by a class discussion about hymen replacement surgery, they'd made appointments with a gynecologist and pretended they were non-virgins on the brink of marriage. Although the doctor initially denied she had ever performed the procedure, under the skillful questioning of my cub reporters, she eventually confessed, filling them in on how much the surgery cost, whether it hurt, and how frequently it was requested.

I had my lead stories. I also had a name and masthead that made me giggle—although that, too, had been a struggle. When my students had put forth their first suggestions—*Our University, Bishkek Today* and *Young Kyrgyzstan*—I'd railed, unwilling to believe that their every spark of youthful creativity had been doused by thirteen years of authoritarian education. Finally, Artyom was inspired: *Slavonic Students' Special Report*.

The name loses everything in the translation, but trust me, it was clever, really clever, in that bilingual context. The English acronym SSSR is the Russian acronym for USSR, so the play of initials was delicious. For the masthead logo, Artyom had scanned the emblem off an old issue of *Pravda*—a tri-flor of the likenesses of Marx, Engels and Lenin—and replaced the old Soviet images with photographs of our rector, our dean and Katsev, our department chair.

The fourth- and fifth-year students were appalled at the irreverence of the design, which I took as a compliment to the younger students, and to me.

"Modern, sassy," I'd declared it, approving the masthead with satisfaction.

I was just back from Chayek and the Aurora when Artyom called to tell me that the university administration had celebrated the distribution of that first issue—and the annual international holiday honoring freedom of the press—by branding the newspaper "fascist pornography," destroying all of the 1,200 copies they could find and banning future publication unless we submitted the articles and a mock-up in advance for official approval.

The skinhead article, with its photograph of a young man with a swastika tattoo, was unacceptable because the writers, heeding good journalistic practice, had failed to condemn such behavior. That was so politically incorrect that the Ministry of State Security U.S., the new name for the old KGB, had demanded a meeting with the authors, seemingly determined to use fascist techniques to prevent the rebirth of fascism.

The hymen replacement story, illustrated with a campy photo of a young woman in a miniskirt and a traditional veil, was not only pornographic but allegedly a stellar example of "yellow journalism." The truth, clearly, had been deemed provocative.

Artyom was the messenger, relaying these criticisms from Katsev, who also sent along his apology that he couldn't call me directly because he spoke no English. Since we'd established a means of communication through Regina, I suspected he'd simply been avoiding my wrath.

"What do you think?" I asked Artyom, waiting, hoping, for some sign of annoyance, if not rage. In a world in which hierarchy becomes an excuse for group passivity, I'd decided to run the paper without an actual editor, but the ever-responsible Artyom had assumed that mantle.

"What is there to think?" he murmured, seeming old, as he usually did.

Prevailing upon my reporter friend Mamasadyk to translate, I made an appointment with Katsev for the following morning. When I walked into the tiny Journalism Department office, I found, to my astonishment, fourteen of my students sitting on desks, crammed into corners and lining the walls. Katsev couldn't stop pacing, his face a dangerous red. He was clearly not pleased to have an uninvited audience. I didn't even try to contain my smile.

No student dared utter a single word as Katsev ranted, I raved and Mamasadyk tried to get a word in edgewise. "How can you call yourself a journalism professor yet require us to submit articles for censorship?" I demanded, leaning forward in one of only two chairs in the room. For once, I had not been offered tea. The secretaries were too cowed by the electricity in the air for niceties.

"Ah, my dear, you're not in America anymore," Katsev declared, dripping with paternalism that needed no translation. There was more, I'm sure, but Mamasadyk found it impossible to follow the full range of parables, quotations and poetic allusions with which Katsev turned the simple into the complex.

"And we're not in the Soviet Union anymore either, Alexander Samuelovich," I retorted, dripping with sarcasm. Within weeks of my arrival, I'd realized that in fighting with Katsev I was operating on my own cultural turf, the battleground of two Russian Jews. Sarcasm and verbal castration were well within the bounds of permissible Jewish argumentative shtick. I'd been trained to it by my father, a master. So had Katsev, by his.

"What lesson are you teaching these students?" I screamed. The phone kept ringing—always for me. Too furious to converse, I kept hanging up on the caller.

Having never seen a Woody Allen movie, my students watched, stunned both by their first glimpse of anyone defying authority and the pitch of the fight.

Gradually, Katsev deflated. I suspect that my bluster and the brilliance of my argument had less to do with his surrender than his fear of the U.S. Embassy, the great granter of funds for equipment and visiting faculty. At that moment, he was still waiting for a response to his request for $16,000 for a television studio.

"Okay," he conceded, finally, "you can continue. But, remember, you must conform to our Kyrgyz journalistic traditions."

Just what my students needed, I thought, yet another professor to teach them the art of reporting by press release, disguising a political agenda as news and self-censorship. Why bring me from New York when there were dozens of old *Pravda* hands in Bishkek who had honed those skills well before I'd entered journalism?

Back in the classroom, defiance tinged the air. Kyrgyz journalistic traditions were not on the table.

"Did you hear him say that he just wanted to preview the next issue for taste?" fumed Alina, a redhead just discovering her fieriness. "What makes him think his taste is better than mine?"

Dinara pulled at her signature baseball cap and grumbled, "This country has no future."

The quiet ones looked uneasy at the energy in the room. "Maybe we should forget it because . . . ," one nearly whispered. Before she could finish her sentence, she was drowned out by ten other voices.

I slouched on the table at the front of the room, grinning. The administration had lit a spark. I had one month left to fan it into flames.

"Should we change the name of the paper and the masthead?" I asked. No one had mentioned either, but when I saw Katsev's face, I heard strains of the "Internationale" ringing in the background. Perhaps mocking Saints Friedrich, Karl and Vladimir had been a tad provocative.

"Can we turn it around?" Artyom suggested.

In its new incarnation, *SSSR* became—well, again, there's no easy explanation in English, since all the best Russian jokes involve word plays. It became *USA Today* in Russia, the Russian acronym standing for "The Student Bureau of Agitation." The new logo was the Statue of Liberty sporting a *kolpak*. American diplomats, I assumed, would show a greater appreciation for irony than had Russian and Kyrgyz administrators. Or at least, I hoped so.

I don't want to make it sound as if the publication of the newspaper sparked a revolution. My graduate students, the fifth-year class, had vanished in the despair of fruitless job searches, and my fourth-year students were so enmeshed in personal turmoil that most were barely coping. Elvira had produced a wonderful short piece about the adventures of a Swiss physician in the arcane world of Kyrgyz medicine. But she had been sick for months and seemed to be willing herself to die rather than face what she and her friends were convinced was a dismal future. Angela's fiancée had been injured in a serious car accident, and his father had vowed not to pay for necessary surgery unless he traded in his Uzbek-Russian girlfriend for a Kyrgyz woman. Rada had not forgotten how to dream and was trying to produce a mock-up of what she planned as the first Kyrgyz women's magazine. But day by day, her hopes seemed to dim as she watched her friends Alexandra and Olga plot their escape from what they called "this hopeless land."

Only the freshmen and sophomores nourished enough optimism for either moral indignation, hard work, or the belief that they could make a difference—to their own lives if not the lives of their countrymen.

Their work remained sloppy, and the copy came in late. But the stories were more solidly reported and slightly less timid. A freshman boy wrote a lovely profile of an eight-year-old beggar who seemed to live on the campus's front plaza. A sophomore girl turned in an exposé of the filthy, malodorous student bathrooms, complete with a photographic comparison between them and the pristine porcelain at the American University. And Dinara produced a grim analysis of the job market for new graduates.

The only piece that might prove inflammatory, I thought, was mine, a guest editorial entitled, "What's Wrong with This Front Page?" Running in both English and Russian, it was the tale of censorship, democracy and the administration of KRSU.

I watched over the copyediting of the bilingual edition like a hawk since the profusion of niggling typos and misspellings had been the single criticism of the first issue that had some merit. While Ahoura, the Iranian exile student from AUK, designed the new issue, Dennis reserved press time at the printer.

Three days before we were scheduled at the printing plant, Katsev stopped me in the hall, casually, as we both wandered between classes.

"When will I see the newspaper?" he asked.

"It will be printed on Monday," I responded. Having established that if my grandparents had remained in Eastern Europe, I might have turned out like him, we were ever so friendly.

"Oh, no, first I must see what you will print," he declared, as if repeating information I had already received. I suspect that he'd sent me that "message" through a student who'd decided not to pass it along, although no one copped to that version of reality. But my back arched at his unilateral withdrawal from our agreement.

"Oh, no, you will not see what we will print," I insisted. "That would allow you prior censorship, and that I will not abide." Regina struggled. Expressions like "prior censorship" and "will not abide" were beyond her vocabulary. But words were not necessary. Katsev could hear my ire echoing down the hallway as I stormed off in a huff.

Katsev—poor Katsev—was caught between another fiery Russian Jew and his bosses, who expected him to keep even the foreign hired help in line. He could stomp, scream, plead and bark orders, but he had no power to contain me, a new concept in a country where everyone followed orders.

At 11 P.M. that night I received a summons, delivered by Katsev's English translator, to appear at the office of the vice rector, Ednan Karabaev, at 10 A.M. the next morning. Katsev and the rector, it seemed, had decided to bring out the big guns. The former minister of foreign affairs, Karabaev was a pacer who wandered the halls of the university carrying an ashtray and a lit cigarette. He spoke more quickly than I did, wore his confidence like a well-tailored suit and was famous as the country's most skillful negotiator.

"I am not available at 10 A.M.," I told Katsev's translator, assuming she would pass along that message.

Regina spoke superb English but was easily cowed by authority. I needed a translator who wouldn't be so easily intimidated and called Konstantin Sudakov, the student who'd challenged the president of the American University over the firing of his boyfriend.

Just after 11 A.M. the next morning, Konstantin phoned Karabaev on my behalf to set up a meeting at a "mutually convenient" time. "Oh, and, by the way, Dr. Burkett would like to know the agenda," he said, following the script I'd provided.

By the time we arrived that afternoon, Karabaev had lost a bit of his bluster.

"You must understand how things are in this country," he said, the sadness heavy in his voice. On my way over to his office, I'd prepared a mental list of the tactics he'd use to try to cow me into subservience and wasn't surprised that he'd begun with a plea for empathy, if not sympathy.

"I understand very well how things are in this country." I staked out my ground. "But that doesn't change my position on prior censorship and freedom of the press."

Karabaev invoked tradition, history, respect for foreign cultures and fear of his superiors. I was a broken record: "We won't submit to censorship." "You can't build a democracy without freedom of the press." "Lighten up, it's just a student newspaper."

In my career as a journalist, I'd battled bureaucrats who tried to keep me from records I had every right to see, politicians averse to prying questions, and dozens of threatened lawsuits. But I'd never experienced a full frontal assault on freedom of the press or heard any attempt to justify it on anything less than life-and-death grounds. The more Karabaev spoke, the clearer I became. The final lesson I would teach my students—perhaps the only lesson worth remembering—had to be about defending principle.

"You're putting us in a very awkward situation with the U.S. Embassy," Karabaev finally pointed out. I'd assumed that the embassy's continuing largesse was his primary concern, but I was surprised at his candor.

"You should have considered that possibility before you demanded that I submit to prior censorship," I responded.

"The embassy can't tell us how to run our university," he declared. His resentment of the strings attached to American money was boiling over.

"That's certainly true," I agreed. "Just as it's true that you're not required to request money from the embassy, and that the embassy isn't required to give it to you. Why should I, as a taxpayer, spend my money to send a journalism professor to Kyrgyzstan if she's expected to teach a type of journalism that runs counter to every-thing Americans believe in? Certainly the American government can't be expected to pay for things it disapproves of."

The conversation was winding down as we both recognized our

mutual implacability. Karabaev rose, claiming the last word as his right.

"You will not publish unless you submit the articles for our inspection," he declared firmly.

I said nothing. The last word is rarely offered verbally.

"Ignore him," Artyom and Dinara counseled. "He doesn't have the right to stop us."

Rights were not the issue, I thought nervously, feeling more like a skittery Kyrgyz than an angry American. I was worried that the university would yank my students' scholarships.

"He can't do anything to us," Artyom insisted. "He wouldn't dare."

The embassy, which was footing the bill, tried a little behind-the-scenes persuasion, which got us nowhere. When I asked the public affairs officer what he thought I should do—"I'm not promising to follow your advice," I warned, "but since you're footing the bill, I thought I'd offer you a chance for some input"—he responded unambiguously, "We at the embassy support freedom of the press." I took that as a signal that they wouldn't exile me to Omaha if I pushed the envelope.

Striking the perfect note was a challenge, especially since I'm of the in-your-face breed who would have mounted the barricades back home. But when you're overseas, no matter how pure your intentions, it's all too easy to look like an Ugly American. Ugliness, after all, even more than beauty, is in the eyes of the beholder, and the beholders in the former Soviet Union were already wary. I assumed that the university administration had already dismissed me as one of those arrogant, know-it-all Americans that I've always despised, the people who hop around the world preaching "the American Way" as the cure-all for everyone else's ills.

So slithering out of the bind wasn't uncomplicated. They'd brought me to Kyrgyzstan to teach a new set of skills, a new way of thinking, but what they really wanted was validation. They wanted me, the "foreign expert," to come and tell them what a wonderful job they were doing. They weren't venal. Like most of us, they were just loath to admit—really admit, to themselves—that they needed to change. Pride doesn't always go before the fall, and even when it does, it doesn't depart without some kicking and screaming.

The old habits of seventy years of communism were dying hard, but each blow I struck raised the specter of backlash, not just against me or against Americans, but against young people just beginning to discern the shape of a different social vision. In the end, of course, they, my students, were the point. I'd promised to teach them everything I could manage about American journalism.

So on a spring afternoon, in the middle of final exams, Dennis, Konstantin, I and my students set ourselves up as newspaper hawkers on a street corner across from the university. I was willing to acknowledge the rector's right to bar the paper from his campus, but I needed to show my students the line between the rights of those in authority and the rights of individuals. In this case, the rights of the former ended at the edge of the university property.

Artyom stashed copies in bathrooms, Dinara dumped them at student hang-outs around town, and others distributed the eight-page edition to every student who walked by. For the first time in their lives, they were actors in a drama of their own writing.

As they pressed papers into the hands of passersby, with more steely determination than elation, the university chancellor drove past in his Zil, a Russian luxury vehicle on the order of a Mercedes, and glared. He didn't bother to slow down. There was nothing he could do since a little bit of defiance is often all it takes to shift the balance of power.

Yet the next year, I knew, no guest lecturer was slated to teach in the journalism department, so no outsider would be around to goad the students. Without that unsettling influence, the balance of power favored those in power.

But as I watched Artyom's face, wide open to endless possibilities, I wondered. Maybe it was just hope. But I let myself believe that the future just might happen.

20

Back to the Future

One day as a tiger is worth a thousand as a sheep.
—UIGHUR PROVERB

Every morning, just before 9 A.M., the hordes descended on the
Russian Embassy in Bishkek, begging for permission to return to
the motherland. Short, fat babushkas jockeyed for space at the gate
with heavily made-up young women in open-toed heels and old
men with World War II medals pinned on their tattered jackets. The
concept of lines had disappeared along with ration cards, just as
personal space had vanished with the introduction of private property. It was dog-eat-dog at the den of the Russian bear.

As they grabbed at the fence, the visa seekers jutted out their
elbows as if declaring to those behind, "Go no farther." But we had
learned, by then, that he who was closest and screamed the loudest
ran the best chance of being admitted to the inner sanctum, especially if he was yelling, *Americanski diplomatika!!!*

I was not, of course, a diplomat, but I had a Kyrgyz diplomatic
card that entitled me to a waiver of VAT charges in stores and
restaurants. Dennis had forged it into a weapon that could get him

past any security guard, any embassy bureaucrat, no matter how obstreperous.

Visas are the bane of travelers who venture beyond the predictable round of European capitals and international resorts, a torment of lines, delays, surly clerks, faux "letters of invitation" and ridiculous fees. Dennis had already conquered the consulates of Afghanistan and Uzbekistan, Turkmenistan, Tajikistan and Iran. When we decided to go home via Russia, China and Vietnam—to meander among America's old enemies in the age of new ones—he was ready.

Getting a visa sounds like such a straightforward process, a minor hassle, like renewing a driver's license in New York City. After all, what government would want to discourage foreigners from spending money in its country? The answer: the Russians. It wasn't that Moscow had adopted an official policy of discouraging the very tourists most countries compete to attract. But after almost a century of Soviet paranoia, and unrestrained by such quaint customs as "customer service," they'd raised bureaucracy to a high art. In a system dedicated to the aggrandizement of petty functionaries, no transaction could be completed without the intervention of at least five clerks and two supervisors.

Dennis's first foray to the consulate, to pick up our visa application forms, had sounded like a simple mission, but it had turned into a three-hour test of endurance as he was shoved, poked and elbowed by the throng while the guard puffed peacefully on his unfiltered cigarettes, studiously ignoring the near-riot. Filling out the forms on the spot was *verboten*, of course. That would have been entirely too user-friendly. It would also have been impossible, since the application bore a striking similarity to the forms used by the FBI for those seeking top-secret security clearances, requiring, for example, full disclosure of all the schools we'd attended since birth. It also had to be accompanied by a "letter of invitation," a conceit that had nothing to do with hospitality and everything to do with lining the pockets of the Russian-based travel agencies that were entirely fictive hosts.

When Dennis returned to the consulate to drop off our sheaves of documents, the gaggle of hefty women, ancient veterans and overdressed ingenues looked suspiciously familiar. Did the same people return, day after day, hoping for Moscow's mercy? Or was their presence yet another sign that Dennis, too, might become a semipermanent fixture at the gate?

Two weeks later, at the appointed day and time, Dennis trudged up Razzakova Street again, pumping himself up for what had been promised as the final assault by humming the theme from *Chariots of Fire*. Within minutes of his waving my magic diplomatic card, a guard had started, as if Putin had appeared on the premises, and glared directly at him. Dennis lofted our passports and a wad of cash optimistically, his pantomime for "I give money and pick up visa." The guard nodded his head, encouraging Dennis's fantasy that his mission was almost accomplished.

"Cum bahk at two-thirty," he said in a deep baritone growl instead.

As he walked back that afternoon, Dennis remained hopeful, intoning his mantra, "This is the game, the test." I remained at home, certain that he would fail.

It wasn't yet two o'clock, but the usual crowd had already gathered, sitting on the bench outside, feeding the suspicion that they were permanent camp followers. Examining the competition, Dennis casually sidled over toward the fence and plunked himself down at its base, just three feet from the gate. The mob tensed. "Is it time to take position? Should we make our move?" Two dozen faces glanced down at watches. "No, it's still early."

Then, a stylish if threadbare couple in their sixties—the woman displaying a dye job that made her hair look fire-singed, her unshaven husband weighed down with medals from the Great War—sauntered by, staring unobtrusively at the ground, as if to suggest that they were headed for the alley. At the last moment they bolted for the gate. In the blink of an eye, the swarm surged forward, tits to ass, shoulder to shoulder, rumbling complaints that the couple had cut to the front. The man swelled his chest to brandish his medals, the woman flashed her passport knowingly, and Dennis grabbed a bar of the fence and held his ground directly behind them.

Before the guard's shoulder had fully emerged from the building, Dennis took full advantage of his height to wave our passports

and yell, "Amerikanski diplomatika!" His grammar was terrible. The
guard didn't care. He ushered Dennis through the courtyard and
into a dark, narrow room with three desks, eight chairs, a metal
detector that had died an ignoble death years earlier and not an
official in sight.

Twenty minutes later, the same short, fat officer who'd taken our
applications appeared and directed Dennis to the window labeled
KACCA, cashier. Busy stamping sheaves of visas—four stamps on
each document—the cashier, a blond woman with thickly penciled
Joan Crawford eyebrows, ignored him. Customers, after all, cannot
be permitted to interrupt serious paperwork. Only after her forms
were neatly stacked and restacked did she deign to look up.

Three weeks and $250 after he'd begun, Dennis came home tri-
umphant—and began planning his sortie on the Chinese consulate.

Leaving is always more difficult than going. The sadness is keener,
the logistics a thousand times more complex. The previous August,
we'd purchased our plane tickets to Bishkek on the Internet, paid
with a credit card and received them by FedEx. The embassy in
Bishkek had arranged our Kyrgyz visas, which we'd picked up in
Washington, D.C., without waiting even five minutes in line. Then
we'd closed up our house, packed three suitcases and departed.

Our trip home required six visas and tickets for eighteen planes,
four trains, six boats, two cars and three buses, none of which were
available on-line. We couldn't even find an electronic schedule of
domestic Russian flights, the breakup of Aeroflot having turned the
Russian airline system into a pastiche of competing, but not connecting, carriers.

We planned to begin our long trek in Russia—not in the chic
restaurants of the urbane New Russians, in Moscow and Saint
Petersburg, but in the old Russia, in the Urals, Bashkortostan and
Siberia. With Artyom as our guide and translator, our first stop
would be Chelyabinsk in the southern Urals, considered by many
to be the most contaminated place on the planet. After a visit with
Artyom's family in a small settlement near Ufa, we'd fly east to
Irkutsk in Siberia. There, we'd catch the Trans-Mongolian Railroad

and wend our way along Lake Baikal and the steppes of Outer Mongolia to Ulan Baator, where I was scheduled to conduct a week of seminars. From Mongolia, we'd take a plane to China and spend five weeks crisscrossing the country by train, plane and bus.

We still fantasized about making it from Beijing to North Korea, but the prospect was fading. A group of Kyrgyz passport clerks had been caught after selling documents to a Vietnamese man, who was then nailed when he tried to enter Turkey on his false papers. The price of an illegal passport had skyrocketed to $5,000. We'd heard that Uzbek passports could be had for a mere $500, but visions of that Vietnamese guy rotting in a Turkish jail because he spoke neither Kyrgyz nor Russian had dimmed our ardor for illegitimate documents.

By late July, we hoped to be in Vietnam and make our way down the coast, then up the Mekong River to Cambodia. After a quick stop in Thailand, we'd end our journey with three weeks in Myanmar, old Burma, a Disneyland of Buddhist temples and military control.

We were braving no new worlds, but those old worlds had been so long, so recently, forbidden territory that they seemed beguilingly new, tantalizingly promising as foils to today's proscribed lands.

Our friends back home were alarmed, anew, at our plans, convinced, like so many Americans, that overseas travel meant diving into a gulf of ill-will. But each time we came up for air after an excursion—a foray to Kabul, a journey across Iran, a visit to Iraq— we were alarmed, too, not by the aura of animosity, but by its absence, by the chasm between what we were reading in the American media and what we were living.

We'd caught plenty of whiffs of anti-Americanism on our travels, but the fires stoking it were rarely set off by the policies and realities being hotly debated back home. People were upset and angry about things Americans couldn't imagine, like the winter Olympics, and exercised not in the least about the encroachment of Hollywood or the Gap. In Central Asia, anti-Americanism was almost vestigial, the persistently acrid smell of the Cold War. And in Iran, much of the old anti-Americanism had been transformed into palpable nostalgia for a long-lost friend.

Even the most virulent revilers of the Stars and Stripes were tangled in contradictions, spouting half-truths based less on U.S. foreign policy than on local political machinations. They were simultaneously applying for visas to a country they claimed to despise, and demanding that America cease its adventurism while insisting, with equal intensity, that America solve the world's problems.

In New York and Washington, Chicago and Dallas, Americans searched for something they could do to prevent another disaster— for some individual or group they could punish or for some new policies and programs that could win hearts and minds. But the most devastating of truths was that much of the welter of emotions and furies behind September 11 had less to do with anything America did than with what America represented—as much our tolerance and our appetite for change as for our materialism and our prosperity. And often it had nothing to do with America at all but was the legacy of British adventurism and almost a century of Soviet domination, of centuries-long tribal hatreds, local political corruption, and internal struggles between modernization and tradition in which Uncle Sam is, at best, a bit player.

By June, with the snow receding from the mountains around Bishkek, we'd glimpsed the outlines of those other forces that felt so impervious to the best or worst of American intentions, and we hoped to shade them in by meeting old enemies in this time of less familiar ones by going behind the Iron Curtain across the Bamboo Curtain into Red China and Vietnam.

I wasn't seeking answers, exactly. In ten weeks I could hardly plumb the depth of the world's love-hate relationship with Uncle Sam. My impulse was less tangible, perhaps more personal. It was my blessing, as an American, that hatred had never been part of the curriculum of my daily life, but it was also my burden that I could not read its script. I'd traipsed the streets of most of America's late-twentieth-century adversaries, from Berlin to Havana, Tehran to Baghdad. But the only place I'd seen open hatred—grimaced proud smirks—was in dusty and desolate refugee camps on the West Bank and in the shops of East Jerusalem, where my parents had lived for two decades. There, people wore their loathing with pride. It was a badge of honor to proclaim: Israel is why we live in misery, so we despise Israel and America, its greatest ally.

In Afghanistan, Iran and Iraq, however, the streets had been an emotional stew spiced more heavily with hospitality than with hatred, with more hope than resentment. And in Central Asia, where the traditions under assault by the West were more cultural than religious, the emotions were even knottier. Wounded pride flared as resentment of the victor whose inexplicable triumph over communism had caught everyone off guard. The keen sense of abandonment and broken promises of lifelong security spilled over as bitterness toward the more fortunate. Envy competed with pique, which was fed by old propaganda and new irresponsible, or politically ordained, reporting. Mostly, the atmosphere was charged with fear—fear of isolation, fear of starvation, fear of feeling small and insignificant, fear of missing the future, yet also of losing the past.

Were those the ingredients that boiled over into terrorism? Maybe, although they hardly made up the recipe being served up by the media. In the wake of September 11, I wanted to take the measure of the streets, to search for the countenance of old hatreds so that I could parse out their ingredients, or perhaps I hoped that by mingling with old adversaries—many, the latest of friends—I could begin to fathom how abhorrence dissolves.

In those final days, Bishkek took a beauty pill, or perhaps it was just that familiarity breeds as much affection as contempt. The crisp early summer light at 2,400 feet above sea level turned even soft colors into rich hues, the leafy parks and tree-lined streets hid the worst of the harsh lines and bleak rectangles, and the markets were beginning to turn into a riot of color as mounds of fruits and vegetables that had been absent for months reappeared. Even the mold on the buildings seemed to glimmer.

Politically, things remained chaotic, the aftermath of the Aksy murders still reverberating. The prime minister had resigned, allowing President Akaev to again shift the blame. The opposition had responded by stepping up its activities, using the U.S. military base as a club to beat up the president. Meanwhile, unemployment was rising by more than 5 percent annually, inflation remained in the double digits and half the population was mired in poverty.

But the coming of the Americans, which is how I thought of the establishment of the U.S. military base, had washed up scores of foreigners in its wake, and with them, money: money for paint; money to replace the worn-out, rickety chairs in cafés; money to bring energy back to Tsum department store. The airport hummed with energy: lightbulbs shined from every socket in the ceiling, the toilets had been cleaned and opened, walk-on ramps now led passengers directly onto international flights. Business at the Pub and the Navigator was booming, not just with military personnel but with the consultants, accountants and civilian contractors who follow them. For the first time in almost a year, I saw actual customers at Roberto Botticelli, the $200-per-pair shoe store on Ala-Too Square.

Despite the city's spiffier veneer, I was haunted by sadness that had little to do with the still-dilapidated buildings and pockmarked streets, the lack of hot water or a government that murdered its own. Those realities had become too normal to notice, as realities, no matter how grim, are wont to. Some days I worried that Russian gloom had infected me. Most of the time, I knew that my melancholy was more than a variation on a Dostoyevskian theme.

Azamat was finishing up another year at university, his endless pastime. At the age of twenty-five, he had nothing else to do since there was no work in his village, no prospect of work in Bishkek, with or without a degree. He was sustained by fantasies of escape, of sneaking into Germany or winning the green card lottery—the Visa Diversity Lottery—and moving legally to the United States.

Although Regina had already left for a summer job in California, thrilled to be going to her second home, I suspected she'd be equally ecstatic to get back to Kyrgyzstan in August. At the age of nineteen, she'd already become a restless binational vagabond, out of place both in America and in Central Asia.

She, at least, had suffered no problems securing her American visa. My friend Ahoura, the Iranian expatriate, had won a full graduate fellowship to Syracuse University but was caught in a security tangle that cast doubt on his plans. Although he and his family had fled Iran because of religious persecution, as an Iranian, he still needed special clearance from the FBI, and it wasn't clear he would get it in time to start school. Where could he go when the only passport he had belonged to a nation he was no longer part of? He had no roots and no future in Kazakhstan, where his parents lived. He couldn't go "home," but he wasn't a resident of Kyrgyzstan, where he lived.

Konstantin was a survivor, but I suspected that he'd pay a high price for taking on the president of the American University over the firing of his boyfriend, and that that price might be his full scholarship, worth $2,600 per year.

Samarbek had run through his full barrel of optimism convincing himself that he'd win a scholarship to study in the United States. When his letter of rejection arrived, he hadn't quite been able to muster the full plunge into denial. He'd been battered once more when his family's fortunes dissolved in an ill-informed venture into capitalism, the Holy Grail of his plan for the future.

His was the reality that nagged at me during those long nights when a welter of sadness and excitement kept me awake. Like most Kyrgyz, he'd accepted, on faith, that his embrace of the new, the American, would be his ticket to the twenty-first century. But like so many of his peers, even of his elders, it never occurred to him that declaring his intention to board that train might not be sufficient. The world they craved—a world of comfort and computers, of travel to New York and Paris, of buying books on the Internet and livable pensions—doesn't coexist comfortably with the world that they knew. It moves to a different rhythm and is ruled by an alternate logic. It does not, cannot, fit into the framework of their old normalcy.

Sure, names like Turkmenbashi, or Saddam, can be glorified on sleek new buildings in blinding neon, just as tradition can be dressed up in Doc Martens. But there's no greater human divide than the chasm between abject fatalists and those who see themselves as architects of their own fates, and few in Central Asia had begun to measure its depths, not to mention to plan the structure that would bridge it.

Fewer still had grasped the price they would pay for doing so. Entry into the modern world isn't gratis, after all. The price of admission is reshaping a thousand assumptions about the role of the government, the responsibility of the individual and the permanence of tradition. Access to all those shiny refrigerators, widescreen televisions, convenient cellphones and smooth roads erodes the very fabric of life that defines members of traditional societies. Yet even the young men and women who drank *kumuz* while listening to Madonna, who wore miniskirts to their families' yurts, who lived in limbo between the past and an unknown future, hadn't begun to grasp that truth. I still couldn't decide whether they would accept it even once it was staring them in the face, erupt in fury at the Hobson's choice or, like so many others in India, Africa, Asia and the Middle East, opt to delude themselves that the "good life" could be grafted onto the old.

Our last night in Bishkek, Mamasadyk and Dinara invited us to a farewell dinner, paid for by Mamasadyk's new job at the local branch of IREX, a nonprofit organization that administers scores of State Department programs and grants. At the age of nineteen, he was earning twice what his father and older sisters had ever brought in, probably ever would.

As we drove to the Little Land Restaurant, I was tired and grumpy from another day of packing, another day of student complaints. Sending packages home was an all-day affair at the Bishkek post office, where the contents of each box had to be carefully surveyed and itemized before the clerk covered it in muslin, sewed it shut by hand, then sealed it with wax. Between trips to the post office, I was fending off students who called or stopped by our apartment to complain about their failing grades. Since failure was impossible in their other classes, they'd ignored my repeated warnings that I would pass no one who did not turn in the assigned work.

"But we must pass," they insisted. "You must allow us to make up the work."

I'd stood my ground, reminding them that I'd been teaching *American* Practical Journalism, knowing that someone—Katsev or the dean—would change the grades I refused to tinker with.

Worn down by a year bucking the local proclivity for acceptance, like a good Kyrgyz, I hadn't railed when the hot water was shut off on May 18 for the annual month-long cleaning of the city system. Instead, I fumed silently at the stupidity of not waiting until August, when the temperature soared above 100, to conduct the repairs.

But I knew that I was displacing, burying my own grief at leaving home—by then, Bishkek surely felt like my home—in annoyance at the freezing cold water that spewed out of my shower or at student attempts to manipulate me. I longed for my own plush American bed, for the familiar nonodors of American supermarkets, for my friends and my dog and my garden. But I cringed at the thought that I would never again see Azamat's beaming smile or that I wouldn't be around to watch Artyom discover optimism. That I'd opened a door for my students but wouldn't have the chance to guide them through it.

My mood picked up when we gathered on the patio of the Little Land, Dennis and me, Dinara, Mamasadyk, his sister Janylai and her fiancé. The restaurant was located on a hill on the outskirts of Bishkek and would have offered patrons a beautiful view of the city if it hadn't been poised directly above a semi-abandoned factory. Bishkekites, however, crowded the tables outside, enjoying the panache of the sunset from the deck of the hotspot they called the Panorama, seemingly impervious to that blight on the horizon.

The evening was a double celebration, of our departure and Mamasadyk's latest coup, securing his father's blessing of Janylai's marriage to a Russian, a topic she'd been afraid to broach on her own. As we took our seats, she'd started to tell me the full story. But when she uttered the dreaded words, "I'm so sick of tradition," Mamasadyk interrupted. "You shouldn't have to hear about tradition to the very end, no?" he asked.

Our waitress interrupted the conversation, seeking drink orders and distributing menus emblazoned with photos of Leonid Brezhnev, the Russians' favorite former Soviet premier. The restaurant was named for *Malaya Zemlya*, "Little Land," the first part of his 1980 trilogy.

I opted for the Brezhnev schnitzel, but, repulsed by the globules of fat that were served whenever he ordered meat, Dennis had

become a poultry-phile. That night, however, no chicken was available, a shortage that had left him eating bread and salad since March,
when the Russian Federation banned the importation of Bush legs.
In firing the opening salvo of the Chicken Wars, President Putin had
spouted high-minded rhetoric about protecting the public from
antibiotic-ridden and salmonella-infested American birds, although
most observers assumed that he was retaliating against Washington's
imposition of a 30 percent tariff on steel imports which threatened
to deprive the Russians of $750 million in revenue. Coincidentally,
that was virtually the same amount the American poultry industry
earned from exporting chicken to Russia.

The Kyrgyz had neither a poultry nor a steel industry to defend,
and in a country rife with unpasteurized milk and uninspected
meat, government concerns about the health risk posed by American poultry rang hollow. But even as they welcomed American
troops to Bishkek, Akaev and his ministers couldn't bring themselves
to ignore Putin's lead and had imposed their own ban. Call it habit.
Call it instinct. Even call it tradition. They'd bowed to the wisdom of
Russia for too long to be unmoved by proclamations from Moscow.

The battle over Bush legs became my final metaphor for Central
Asia, and beyond, for the struggle over identity, ideology and imperial control, for the contest between tradition and change and the
abyss between hope and powerlessness. I admit that as I wandered
the markets and bazaars in my final weeks in the city, listening to
consumers bitch and moan about the high price and low quality of
local chickens and to a few dissident voices whispering rumors
about American imperial plots to poison Russians and Kyrgyz with
toxic fowl, I indulged myself in full metaphor-building.

Then, on a rainy spring afternoon when the city seemed particularly steeped in its own dreariness, I overheard two elderly
women, their faces heavily lined maps of hard lives, standing in
front of the meager selection of poultry at my local market.

"I don't care whether this is about steel tariffs or antibiotics,
about competition between America and Russia for world domination, about whether Akaev stays president or the war on terrorism. Can't anything in this country be about what it is? Can't they
just leave me alone so that I can afford to buy chicken?"

Epilogue

The legend is still fresh but hard to believe.
—RUSSIAN PROVERB

"It is easy to see the beginnings of things, and harder to see the ends"; at least, that's what Joan Didion wrote, in a long-ago essay of her time as a young woman in New York. For years I was irritated by that musing, convinced that beginnings got hopelessly tangled in middles and mucked up by finales. But as I leaned my shoulder into the door back into the neutral zone between Bishkek and the plane to Russia, my first moments in Kyrgyzstan retained a crisp clarity, not just of the dim lights and crush of taxicab drivers at the airport, but of my own breathlessly optimistic certainty that I could grasp Central Asia in a single, year-long gulp. And although I was living it, the ending indeed felt fuzzy, as if I were a player in a drama with a denouement still unscripted.

Somewhere between the high mountains of Central Asia and the vast steppes of Russia, I began to suspect that a comfortingly tidy windup eluded me because I'd been mugged by Aesop at a formative age. Even in this self-consciously postmodern era, when old-fashioned fables seem as corny as Mickey Rooney films, I couldn't build an ending, written or lived, without the promise of a moral to the story. Yet I distrusted such morals, even when they were my own.

As the boundlessness of lethargic Russia built to the boundless frenzy of China, and as the dynamism of her billion citizens all shopping at once softened in the green haze of Indochina, however, the strands began to weave together, and a pattern, or maybe really a slightly messier tapestry, emerged and took hold. I carried it back with me across the time zones and borders—across a vast distance that was so much more than geographical, as the world events during my year away had vividly shown—and slouched toward home.

At the new mall in Ufa, a city of more than a million on the western slope of the Urals, well-heeled shoppers shelled out $100 worth of rubles for cellphones, and then drove home to apartment complexes where the garages had been converted into pigsties. Punk rock wannabes hung out by the snack bar flaunting FUCK LENIN and EAT THE RICH T-shirts although their fathers were paid in scrip, not cash, by the factories where they worked.

In Chelyabinsk, in southwestern Siberia, lines formed early at the new Baskin-Robbins, although at the Victoria Hotel down the street, the ubiquitous floor ladies zealously guarded the toilet paper, doling out the precious commodity a meter at a time.

And in a wooden cabin on a collective farm in rural Bashkortostan, an old woman served borscht and homemade bread to relatives from Moscow, then leaned back in a wooden chair and sang traditional folk songs in a voice that would have wowed Ethel Merman. The Muscovites shuffled in discomfort, their teenage daughter rolling her eyes and whispering to her mother.

The New Russians bore a striking resemblance to the old, who built their gleaming capital and left their nation behind.

Reeling from the tangle of conflicting images, a flip book that refused to track smoothly from one picture to the next, I reminded myself of the distance between rural Mississippi and Madison Avenue, or between Paris's Île Saint-Louis and Corsica, Europe's own third world. But it didn't work. In the United States and Europe, that expanse is measured in decades. In Russia, it spans centuries, and like many Iraqis, Afghans or Kyrgyz, the people still aren't certain whether they want to bridge it, whether they want to look forward or take refuge in the past.

Like novels, great or schlocky, presidential campaigns and Broadway musicals, societies are built around their national stories. The arc of Russia's—a tale of imperial might, manifest destiny and intellectual greatness—was smashed when communism collapsed, the Soviet Union dissolved and the plummeting of the ruble turned a lifetime of savings into an evening's supper. Most of the country was careening backward, a reality I could fathom only by imagining how I would feel if every major American corporation fell into simultaneous bankruptcy, the states ceased to be united and every penny I'd put away toward retirement couldn't sustain me for a month.

My imagination failed me, which was part of my interest in exploring Russia. What story did the Russians have left to impel them forward, or were they surviving the social earthquake on old illusions?

Artyom's mother, Marina, chortled at the question. "There are no illusions left," she answered dryly, running her fingers through her blond hair. That was closest she came to showing any anxiety. Anxiety, I'd learned by then, had become too normal a state in Russia to provoke much overt nervousness.

We'd been talking nonstop for three days, seated in the tiny kitchen from which Marina served an endless variety of cakes, salads and homegrown produce brought over by neighbors stunned that Americans would actually want to hang out in Chismy, population 22,000. "We read that American tourists sometimes stayed with families in Moscow, and that seemed so strange to us," said Artyom's father, Vladimir, a Russian Orthodox priest. "We couldn't understand why Americans would want to stay in cramped Russian apartments instead of grand hotels. And we certainly never imagined any American would want to come to a place like Chismy."

Russia was the only place I'd ever been where my nationality conferred on me celebrity status. At the annual harvest festival in

Chismy, the provincial poobahs urged us to join "honored guests" for a group photograph, and we couldn't so much as sit outside Artyom's family apartment without being showered with small gifts. Learning that we were Americans, a taxi driver went so far as to refuse payment and ask for our autographs.

On the train between Chelyabinsk and Ufa, we couldn't have slept even if two conductors hadn't been firmly planted in our reserved seats because word traveled up and down the train, along with bottles of vodka, that *Americanski* were on board.

"Can we talk to them?" passengers asked Artyom, begging him to translate.

"Where are they from?"

"What are they doing in the Urals?"

The conductresses, stout middle-aged women who'd worked the rails for decades, quizzed us about American salaries and television, the structure of the government, the contents of the grocery stores and the conditions of our trains.

"Do you see Russian movies in America?" one finally inquired. I suspected she knew the answer but was hoping I might offer her some shred out of which she could construct a lifeline for her Russian dignity.

Only in Chelyabinsk was celebrity awe laced with a dollop of reserve. The center of Soviet nuclear and weapons production, Chelyabinsk had long been a closed province from which outsiders were barred to keep prying eyes away from Soviet nuclear facilities and weapons factories. Most of the old restrictions had been lifted, but many roads remained closed and a handful of villages still appeared on no map. I'd heard of Chelyabinsk only because Gary Powers, a U.S. Air Force captain, had been shot down during a spy flight over the region's military installations when I was in the eighth grade.

The province was a study in ecological cataclysm. Between 1949 and 1956, radioactive liquid wastes from Mayak, where the Soviets produced the plutonium, were dumped into the Techa-Iset-Tobol river system, the source of drinking water for dozens of villages. In 1957 a nuclear waste storage tank exploded, spreading plutonium and strontium over an area the size of New Jersey. And in 1967 a cloud of radioactive dust blew across the region when a drought dried up two natural lakes served as nuclear dumping basins.

Moscow's post-Soviet government acknowledged that more than 20,000 people had been exposed to high levels of radiation and that scores had already died. Leukemia rates were soaring. Infertility was endemic. But Mayak remained the storage site for the spent fuel from almost 250 Russian nuclear submarines, and the Duma, the Russian parliament, had plans to turn the old military secret into a profit-making center as a fast breeder recycling plant for foreign spent plutonium.

The locals took their notoriety in stride. According to polls and surveys, half were unaware that Chelyabinsk was considered to be the most radioactively contaminated territory in the world.

"Everything is clean now, see?" bragged Sergei, the driver we'd hired to take us on a tour of the region's environmental disasters. "The radiation settles on the bottom of the river, so you can even swim here," he exclaimed, pointing down from a bridge over the Techa. "And look at all the green. That shows that everything is all right."

Everything was most decidedly not all right. Some workers at Mayak had been exposed to doses of radiation comparable to those received by survivors of the bombings of Hiroshima and Nagasaki, according to one member of the U.S. Nuclear Regulatory Commission. The banks of the rivers still contained high levels of cesium and strontium, since natural deactivation takes centuries. And fish in one provincial reservoir were reported to be 100 times more radioactive than normal.

"He seemed optimistic," I told Marina and Vladimir. They knew from Artyom that I'd been searching for a Russian optimist and were quick to disabuse my chimera that I'd found one.

"Everyone knows how much damage was done to the environment in Soviet times," they insisted. "He's just decided to believe the government because it's easier.

"You have to understand that you have two choices in Russia: you can believe the official story or you can believe nothing, and the official story is always easier. Oh, some people like unofficial stories, like the one about the Americans paying off Gorbachev to destroy the Soviet Union, or about the glorious days of the czar. And there are always politicians who try to build followings around those stories.

"Some people like stories that let them blame foreigners or the government, but most people blame no one and everyone. They just think, It's just our destiny.

"Is it the same in America, and in other places? We don't really know, because we don't meet many foreigners and because our media only tells us what the government wants us to know. If they don't follow instructions, they're closed down."

I interrupted, as much to give Artyom a rest from translation as to offer an observation.

"That's its own story, isn't it, that the people are powerless to control their own fate?" I suggested.

Vladimir laughed, and his dark eyes sparkled. With the long beard of an Orthodox priest, he could look immensely severe. But he was too curious, too passionate, for severity. "You're such an American, so happy and naive."

All year I'd longed to discuss the backwardness of the former Soviet Union with Russians, to talk about the concrete that crumbled within months of its pouring and the absence of conveniences like brooms or decent refrigerators, about the popular passivity and the presence of outhouses a stone's throw from the seat of government in the capital of a Soviet Socialist Republic. Silly me, I'd assumed that a country capable of producing missiles could manufacture flush toilets.

Things weren't all that different inside Russia, the same shoddy infrastructure, the same lack of concern for consumer niceties. Had Russia, the country Americans had feared for so long, always been a paper tiger? Would the concrete of the missile silos have disintegrated if the thrusters had ever been fired? Had inefficiency and indifference been the theme of daily life for half a century? Did Russians *know* what their country looked like to a visitor from the first world?

On our last night in Marina's kitchen I cleared my throat and dove right in. "I was surprised to discover how backward everything seemed," I confessed. Before I had a chance to soften my words, Marina broke in.

"If you were surprised, imagine how we felt," she said, laughter and anger competing in her voice. "We thought we were the most advanced nation on earth. Then everything fell apart, and we saw America."

Feeling her devastation, I changed directions. "So in this house there are only pessimists?"

Vladimir hesitated then pronounced, "We're not pessimists, we're fatalists."

Marina shook her head in disagreement. "If we were pessimists, we'd commit suicide."

By my twenty-second hour on the Trans-Mongolian Railroad, I felt the first inklings of something akin to euphoria, an odd sensation, since I'd been oblivious to my own melancholia, or perhaps it had just seemed normal in the bleakness of Bishkek. With each mile traversed by the old train, still decorated with a Red Star like a prop in David Lean's version of *Doctor Zhivago,* the sensation, grew into relief. I was tired of the former Soviet Union, weighed down by a year of pessimism or fatalism, call it what you will, of doom and gloom and drab cookie-cutter apartment buildings. Ten miles north of Ulan Baator, Dennis and I packed up our backpacks and leaned out of the doorway, hungry for our first glimpse of the non-Soviet city, a city I'd been dreaming of for fifteen years.

I almost cried when I saw the same tedious concrete landscape that blighted most of the cities between Krakow and Beijing.

Fortunately, Ulan Baator was deceiving. The city's dour design was broken by the occasional yurt, or the occasional neighborhood of yurts that sprung up when nomads from the steppes fled drought for the capital, their houses in tow. Lenin peered down from his perch in a park in front of the Ulan Baator Hotel. But the country's hot designers held fashion shows around the corner, displaying a dizzying array of avant-garde boots and skimpy faux traditional dresses that would have elicited hearty bravos on runways in Milan.

In a city giddy for the new, the old staple diet of mutton, fat and mutton-fat soup had given way to a culinary riot of French bistros where Mongolian waitresses offered up Beaujolais, *s'il vous plaît*, of Italian restaurants decorated with wolf skins and yak horns, and of Japanese steakhouses done up Texas-style. Got a yearning for rice and black beans *à la cubana* after a long day riding a horse across the steppes? You could order a hefty serving at El Latino on Peace Avenue across from the bakery/Internet café. Prefer bacon and eggs? *Pad thai?* Curry? You could even take a meeting over lunch, along with three-quarters of the burgeoning ex-pat population, at Millie's, run by a Jamaican from Berkeley.

UB, as the in-crowd called it, was a riot of BMWs streaming past barefoot Buddhist monks, of Soviet-era trolleys, satellite phones, a stock exchange inside an old movie theater with a trading system designed by Harvard grads and old men wearing hand-tooled leather boots with felt linings and woolen coats called *dels*.

I was in such awe, and in such need of good food, that it took me five days before I realized that every conversation I had included a complaint about China: Chinese companies are trying to buy up Mongolia; Chinese men are marrying Mongolian women as part of a plan to capture the country for themselves; the Chinese government is plotting invasion; the Chinese aren't as good as the Russians.

"The Chinese are dangerous neighbors, always attacking us," the Mongolians declared. The last attack had occurred in 1921.

Every schoolchild could recite a long litany of atrocities meted out against the Mongolians by the Chinese, who were guilty of ethnic cleansing, "sinofication" of the 4.8 million Mongolians living in China, the persecution of Mongolian intellectuals, holding Inner Mongolia captive and trying to usurp Genghis Khan as one of their own.

The accounts had the familiar ring of so much of "official history": In Mongolia, the Chinese were awful but the Russians were great. In Syria, everything was wonderful until the Jews created Israel. The British are to blame for all Irish problems. Europe is the center of civilization straining to hold back the barbaric Yankees. America has always been a beacon of hope for the world.

In what I think of as my first life, I was a historian, a university professor of history, and nothing raises the hackles of a historian more instantly, more thoroughly, than politically inspired historical revisionism. Begin your version of history in 1900, and the Europeans look like imperialist swine; start the story a century later, and they are the paragons of tolerance. Look at a textbook from Palestine in 1905, and the Turks, not the Jews, are the menace. Ask a Soviet historian about the Cold War, and you'll hear an almost unrecognizable tale. Talk to my students in Bishkek, and you'll learn that neither Kuwait, Somalia nor Bosnia counts because the United States has never supported a Muslim nation.

The good guys and bad guys trade roles so facilely that history becomes a mix-and-match ensemble that can be endlessly rewritten. Don't like your national story? No problem, you can adjust it to blame someone else, at least if you don't mind shading a few facts or truncating whole centuries. Want to seize power from the opposition? No sweat. The past can be turned to many uses. Is your country's autobiography in conflict with someone else's? Don't worry. Burnish off its edges and call them liars.

Does this sound cynical? Sure it does. But history has always been a cynical political game in which all sides pretend that only the others mangle the truth. Selectivity becomes reality because power mongers and zealots call the shots, and all the "ordinary people" swallow whichever version fills their bellies or their sense of pride.

"Why is everything China's fault?" I asked my students once my stomach was full and my patience depleted. In Mongolia, they were mostly professionals who'd signed up for my seminar at the Mongolian Press Institute. They instantly began spewing the official line about Chinese imperialism, Chinese invasions of Mongolia, Chinese wreaking havoc on the peaceful Mongolians. Having spent the preceding week reviewing everything I knew about the centuries of misery the Chinese had endured at the hands of the Mongolian hordes that descended from the north to pillage their cities and farms, I finally interrupted:

"Why did the Chinese build the Great Wall?"

After ten days in Mongolia, we crossed the Wall and, after being stuck so long in the past, arrived back to the future. China was a delirium of energy and commerce, a raucous symphony of vitality and optimism, in stark counterpoint to the somniferous lullaby of Kyrgyzstan and Russia. The old work groups were becoming corporations and had turned downtown Shanghai into an architect's wonderland of postmodern and post-postmodern skyscrapers, all built simultaneously by carpenters climbing bamboo scaffolding. The 600th Kentucky Fried Chicken franchise had just opened in the northwestern province of Gansu, and you could order up a latte grande or a Frappuccino at the Starbuck's inside the Forbidden City.

Kyrgyzstan had abolished communism, but even the most educated young people didn't comprehend that factories could not reopen without a reasonable prospect that they could turn a profit. The Russians, the planet's latest converts to capitalism, hadn't mastered the intricacies of accounting, airplane schedules or computerized record keeping. But the Chinese were flinging themselves headlong into modernization, studying English, mastering computers, taking driving lessons, plotting investment strategies, luring foreign investment—and the anti-globalization firebrands be damned.

The first time Dennis was in China, the country was a sea of blue Mao suits. In the intervening decade, all one billion Chinese had purchased new Western wardrobes, and the economy was booming. Even in the countryside, peasant farmers kept meticulously and intensively cultivated garden plots and hung television antennae on the roofs of their mud homes. It was tough to find much communism in Red China.

As I stood in front of the immense statue of Mao that graced the Bund in Shanghai, just down the road from where his wife lived during the Cultural Revolution, I wondered what the old guy would think of the new skyline that changed by the hour, of the cellphones and foreign banks, the hookers and the Hyatts. The only vestiges of the Cultural Revolution were flaking copies of the *Little Red Book* and Gang of Four statues on sale at inflated prices in tourist-oriented antique stores.

Beijing's old *hutong* neighborhoods were being razed at a record clip, replaced by towering high-rises that looked just like the towering high-rises of all world cities—until you noticed the pagodas on top. The New Chinese, the Global Chinese who think that internationalism refers to business, not to Communism, were buying $300,000 homes in developments with names like Orchid Villas and Golfing Haven, withdrawing their cash from ATMs and dropping $100 on dinner without blinking an eye. Foreign businessmen—investors, consultants, corporate honchos and their attorneys—flew in by the planeload and were ceremoniously shuttled to glittering five-star hotels.

When we trudged out of our hotel in the center of Urumqi in the far west of the country early one morning, workers were already swarming over the frame of a new office building; when we passed by again at midnight, floodlights had been set up and they were still at it. With the number of newly registered automobiles up 43 percent over the previous year, bicycles were being edged off Beijing's roads. A survey of China's urban households suggested that 26 million families were poised to become car owners, which would dump the entire annual global automotive output on that nation's urban roadways.

The Chinese panted in breathlessly proud wonderment at their own accomplishments: Look, we're finishing a 2,500-mile pipeline from the far West to Shanghai a year early! Are you aware that 73 percent of the young people in Beijing know how to use computers? That the Chinese rank second in the world as mobile phone users? That our gross domestic product grew by 2,400 percent over the last two decades? That we're planting 12 million trees to clean up Beijing's air for the Olympics? Moving 200 factories out of the city?

The energy of fifty years had been uncorked and the Chinese were racing for the finish line: a first-world economy and a Western standard of living within a decade. It was impossible to be a disbeliever.

Yet China remained a deeply traditional nation, or perhaps the word *yet* is misplaced; the new is being built so firmly on the foundation of the old. Mao taught, "Make the past serve the present," and the Chinese remained obedient to that maxim.

In Russia and Kyrgyzstan, tradition was an excuse: Oh, we can't move ahead because we've always been nomads, said the Kyrgyz. We're just lazy. The Russians fell back on their fatalism and their moroseness. But since no body of myth, belief and ritual can survive the centuries unscathed, tradition is invariably selective and open to popular interpretation. The Kyrgyz could say: As nomads, we've always been willing to move and travel in order to have a good life. The Russians could tell themselves that they're survivors who've already coped with so much. Instead, they'd chosen social narratives that left them moving back even as their neighbor China zoomed into the twenty-first century.

The Chinese remained committed to family, to hierarchy, to Confucian ritual, order and obedience. No one dared walk on the grass. Students dutifully wore their school uniforms unadorned. The young were invariably courteous and attentive to the elderly. And in the Shanghai subway, the sign warning you not to jump off the platform tellingly didn't read, "Danger," but "Jumping from the platform is prohibited."

But the orders coming from above were to be fruitful and multiply the gross national product, and the people were obeying without a wit of cynicism, even about the excesses of the past.

"What do people think about Mao?" I asked everyone who seemed interested in conversing, which was every person I met.

"He got it half right," most responded, refusing to break fully with the past. No one in China was suing for compensation for injuries done them during the Cultural Revolution.

"It's hard to understand exactly what Mao was thinking, but maybe without his policies we wouldn't be where we are today," they responded without sheepishness, just with gentle understanding—and a firm conviction that their leaders had their best interests at heart.

The Chinese seemed unfazed by the dizzying pace of change, as if it were, in Kyrgyz parlance, *normalna*. At lunch time, workers lazed by half-constructed high-rises, eating and napping. In offices, clerks signed off their computers and played cards over their noodles. On Sundays, they strolled their neighborhoods, the men in undershirts and Bermuda shorts, or in pajamas, the women in housedresses straight out of the closets of housewives in Nebraska. One steamy Sunday evening in Shanghai, a group of thirty neighbors—young, old and middle-aged—gathered in front of a statue of Bach across the street from the Holiday Inn to dance to the tinny Chinese pop music that brayed from an old tape recorder. Men and women, women and women, followed fixed steps no matter the tune or beat. If the music was slow, they patterned the tango. If it was fast, they parodied a low-key jitterbug.

"Things are really good," an elderly linguist told us. "And we thank America for it," he added, inexplicably.

Things were so good that in the Forbidden City and downtown Shanghai, on the Great Wall and in minority villages like Dali and Lijiang, scores of buses disgorged hundreds of Chinese tourists who were overwhelming the foreigners who'd once threatened to overtake China's dazzling sites. One afternoon, I was sitting in a café in Dali when a contingent of two dozen wandered down the street, each member wearing the group hat, following a guide with a flag.

I looked up and was startled to see fifty cameras aimed in my direction, the presence of a foreign redhead—not so long ago, a foreign devil—having become the hour's attraction.

Suddenly I recalled the rest of the quote from Mao about making the past serve the present: "Make foreign things serve China and let a hundred flowers bloom."

I'd lugged little emotional baggage along on our trek across the vastness of China. Like all Americans of my generation, I'd been brought up with a full measure of low-key hysteria about the scheming Red Chinese devils. But it had taken even less root in my consciousness than Elvis, hula hoops and the rest of my civics lessons.

But Indochina was the leitmotif of my youth. So when I walked into the strangely vacant mansion that was the Vietnamese consulate in Ulan Baator—an oversize white chateau where a family cooking vegetables on a wok atop an electric hot plate processed our visa applications—Vietnam loomed too large in my mind to be

a simple country of peasants eking rice out of hillsides, of fishermen fleeing the flooding of the Mekong River, women donning
long purple gloves to keep their hands white and men weaving
motorbikes in and out of Hanoi's traffic, their hands permanently
glued to their tenor horns.

I'd been imagining Vietnam for more than half my life, although
for most of that time I didn't know I was engaged in a self-deluding
exercise in inference, fantasy, presumption and projection that I
thought of as knowing. Having watched the nightly military trave-
logue of Vietnam for a decade, studied Indochinese history, read
hundreds of newspaper articles, and listened to scores of speeches
delivered by Southeast Asia experts, I believed I'd come close to
that much-maligned, ever-elusive state called the truth.

Like so many of my peers, then, I grew up obsessed with that
imaginary Vietnam. It was the land where old friends unable or
unwilling to buck "the system" were destroyed, by bullets or bull-
shit; a world of faceless, nameless victims of American innocence,
ignorance and greed; the country that robbed me of illusion.

It was my first, perhaps my finest, teacher.

I wasn't the only traveler who brought too much history to the
Vietnamese table. Vietnam had become a sort of pilgrimage site for
Americans of my generation, for war veterans who spilt their blood
and innocence on those amazingly green hills and for antiwar veterans who spilled their hearts. Some went to expiate their guilt,
most just to remember.

I remembered Jane Fonda when our air-conditioned train
pulled into the old Hanoi train station, a flashback too flat to be
either admirable or abhorrent. Everything in Vietnam felt like a
memory. Thankfully, the Vietnamese didn't allow me to dwell too
long in yesterday.

Driving up Highway 1 to the DMZ, I began superimposing
grainy old black-and-white images of smoke and bomb craters on
today's fertile rice paddies. But at a toll plaza, the ticket taker stared
into our car and said cheerily to our guide, "Oh, big people." And
when we stopped at an entrance to the Ho Chi Minh Trail, villagers talked not about how that 20,000-mile stretch of paths and
tracks had fed the war with guns, bombs and ammunition, but

about its new function as a smuggling route for young men bringing drugs and cut-rate cigarettes across Laos from Thailand.

We spent a long afternoon wandering the old U.S. combat base
at Khe Sanh, where American troops endured a brutal seventy-
seven-day siege in early 1968. The scorched earth I recalled from
the evening news had been reclaimed by coffee growers, although
the outline of the old airfield had been maintained and the grounds
around it were still littered with bullets and rusting shell casings.
Inside a small museum, visitors were invited to leave their musings
in a guest book. French and Germans had waxed eloquent about
their solidarity with the brave Vietnamese peasants. The American
vets who'd returned to remember the blitzkrieg of mortar fire,
machine guns and bombing, were more personal:

"Happy Birthday, Bobby, who died on this rotten hill on February 16, 1968."

"A large part of my heart still resides in the red clay of this
place."

"I'm glad to see the coffee growing and the American war fading in the past, but one should never forget the evils of the war."

"I'm sorry. God, I'm so sorry."

But outside, a dozen young Vietnamese, too young to remember
the Tet offensive, sold U.S. military dog tags that would surely have
been the most gruesome of all souvenirs if they hadn't been fakes.
Or perhaps the fact that they were, and that European tourists
pawed through them with the same casual shopper's mien they
adopted when haggling for sarongs while American vets looked on
in frozen horror, made the scene even more mind-numbingly
modern.

In Hanoi, a city of bottomlessly seedy colonial charm—of beautiful old buildings, riotous bougainvillea and tiny shrines inside
bakeries and grocery stores—we went looking for Ho Chi Minh
amid the cacophony of motorcycles, designer motorbikes, bicycles,
old men pushing carts, women balancing heavy loads perched on
bamboo sticks on their backs, and the occasional car. We found
him—literally—at his mausoleum on Ba Dinh Square, where
scores of Vietnamese lined up on the red carpet leading to the
glass-framed sarcophagus where he rests in a plain tunic and san-

dals. Elsewhere, however, Ho was curiously absent. Or perhaps that
absence seemed curious only because Ho Chi Minh had loomed
so large in *my* mind, or because I'd become accustomed to leaders
who inflicted their narcissistic personality disorders on their people.

Even the mausoleum was more than Uncle Ho, as people still
called him, had desired. In his will, he'd requested that his ashes be
buried on hilltops in three parts of the country, saying, "Not only is
cremation good from the point of view of hygiene, but it also saves
farmland."

No matter where we traveled, I couldn't keep my eyes off the
faces of the people, the individual people—the rice farmers trudg-
ing to their fields, the wizened old men in the markets, the women
bent from a lifetime of backbreaking labor. Perhaps I was still
caught in the grips of a youthful romanticism about Ho Chi
Minh's beleaguered peasants, or maybe my sentiments were tinged
with nostalgia for a time when right and wrong, in war, as in so
many other matters of state, seemed clear-cut. But I was moved by
their shy politeness. Even in the harsh light of the vicious Indochinese summer sun, the Vietnamese seemed so lacking in aggression
that I couldn't quite reconcile their public bearing with the grit of
decades of resistance.

In Iraq, I'd been unable to stop thinking about what would be
destroyed; in Vietnam, I was searched for signs of what had been
rebuilt. But once I forgot the past and honed in on the present, the
grit of the Vietnamese became stunningly clear. Between 1987 and
2000, Vietnam doubled its production of rice, transforming their
country from a rice importer into the world's second-largest rice
exporter. Between 1993 and 1996 alone, more than 13 million
people had been lifted out of poverty. And few of the scars of the
war—bombed-out bridges and pockmarked highways and villages
crumbled in ashes—were visible, a public facelift that seemed to
say, "Don't dwell too much on the past, look to the future."

Tourism had become a linchpin of that future, already accounting for almost 20 percent of Vietnam's gross national product. And
while Australians lounged on the beach at Nha Trang and French
swarmed over the weekly markets at Sapa and Khao Vai, the government in Hanoi wasn't naive about the odd allure of war

tourism. Foreigners were invited to slog across rice paddies, crawl on their bellies through old Viet Minh tunnels and race up the Mekong in converted patrol boats—adventures once so assiduously avoided, now paid for up front in American dollars. It's the winner's privilege, of course, perhaps the modern spoils of war.

From the old imperial city of Hue, one of the centers of the 1968 Tet offensive that cooled American ardor for war in Southeast Asia, we took a standard day trip into the heart of the battlefields. With travel permit in hand—$10 per person—we drove up Highway 1 to the demilitarized zone, with stops at the old bases of Khe Sanh, Camp Carroll and Con Thien. After lunch, our guide led us to Vinh Moc village, whose residents protected themselves from U.S. carpet bombing by building a long chain of tunnels that included family bedrooms, a communal conference room, even a delivery room where seventeen babies were born.

Most of the tourists who took that long march were Europeans expressing their solidarity with the struggle of the Vietnamese people against the United States. Curiously, few of them stopped at Dien Bien Phu, where the Vietnamese people's army captured the entire command of the French expeditionary corps, ending almost a century of European colonialism in the country.

In Hue, a city I remembered all too vividly, I was haunted, again and again, by what I couldn't see, by what I could only imagine. Sitting comfortably on a padded pink chair while a Vietnamese man pedaled me through the city streets, past the old citadel, along the Perfume River, where an old man without legs handbiked himself down the road, I wondered, with more curiosity than guilt, "Do they hate me? Does that legless man blame me for his infirmity?"

You can only get answers if you are willing to ask questions, of course, and in Vietnam, I'd been uncharacteristically reticent in posing them. Was I afraid of what I might hear? I wondered. The next day, we hired Hai, a forty-five-year-old English-speaking guide, to show us the villages along the DMZ. Biting back my hesitation, I asked, "How does it feel to take tourists to these villages?"

Hai glanced out of the window of the car, turning in on herself for a moment. "We don't forget, but we try to ignore," she said quietly. "I've been doing this for eight years, and at first, when people realized the people I was with were American tourists, I was uncomfortable. Now, I'm used to it."

She wasn't quite telling the truth, either to herself or to me. Her continuing discomfiture was too patent to conceal. It was Martyrs Day, the Vietnamese equivalent of Memorial Day, and buses filled with veterans of the North Vietnamese Army and the Viet Minh were following our same route. There were no barbeques or football games in the local villages. The war was too recent, the pain too keen.

"I think about the women from the North who volunteered to work along the Ho Chi Minh trail," Hai continued. "They got poisoned with Agent Orange and grew old alone. They lost their youth and, for what and why?"

Her bitterness was almost comforting, reassurance that my imagination—my projection—hadn't failed me.

But three days later, in Hanoi, I posed a similar question to a lively twenty-one-year-old clerk in an antique store on Hang Khay Trang Tien. "I don't remember the war, of course, but I learned about it in school: first the war with the Chinese, then with the French, finally with the Americans. I deal with all people here, and the French I definitely don't like. They treat us like we're colonials and they still run things. And the Chinese, well, they're sneaky. They sneak their people and military across the border and spy on us. And my family says that the Russians were terrible, boring and unpleasant. But the Americans, I love the Americans. They're such nice people, so open and friendly."

There's an almost mesmeric allure to Phnom Penh, Cambodia, part the end-of-the-earth romance of lazing with the geckos under the clattering ceiling fans at the Foreign Correspondents Club while waiting for Humphrey Bogart to show up in a white suit, part the bat-infested French terra-cotta-colored roofs of crumbling colonial villas whose imperialist origins have been softened by years of decay. All set against the unspoken modern seductiveness of a city still shell-shocked by its own dismemberment, by the ubiquitous specter of butchery.

Those charms were lost on most Americans, who headed straight from Vietnam to Angkor Wat, if they braved Cambodia at all. But Phnom Penh was a magnet for a peculiar breed of European who fed on the decaying colonialism and the stark reminders of man's inhumanity to man. By day, they were staples of the new outdoor café scene on Sisowath Quay along the Tonle Sap River, by night, they hung out at Sharkey's Bar in their uniform of DANGER!!! MINES!!! T-shirts, lest anyone confuse them with people who didn't "care."

I'd never considered myself a Europhobe, but by the time we reached Cambodia, I was sick to death of the whiny French and the arrogant Brits. Europeans, it seemed, at least the Europeans who spoke enough English and had enough spare Euros to travel to distant places, had grown bored with September 11 and felt no compunction about venting their pique on random Americans. I, in turn, was bored with Europeans.

In Lijiang, China, a German couple had warned us, with the knowing wink of the "we've been there, we know" crowd, that Dick Cheney was fostering the imposition of fascism on America. In Dali, a young Frenchman told us, with almost parental condescension, "You can't avoid the truth that America brought the attack on itself. Americans act like cowboys. They really need to follow the lead of the EU."

In Sapa, Italians insisted that the United States was creating "problems" in Palestine to distract the world from its preparations for the war on Iraq. And on the two-day boat trip up the Mekong River from Ho Chi Minh City to Cambodia, I heard from a Dutch woman that Bush was cooking up trouble with North Korea to prevent reunification with the South. Then, in Phnom Penh, I wandered into the Heart of Darkness Bar only to receive a lecture from a mixed group of EUers—or are they EUians or EUish?—about the greed of American corporations, delivered, without a trace of irony, inside a drinking establishment named after Joseph Conrad's novel about Belgian imperialism and genocide in the Congo.

I'd already endured a year in which British journalists in Kabul had berated me, as if I were the U.S. commander in chief, for America's "unenlightened" policy in Afghanistan. A Finn in Mongolia had conjectured that the U.S. government was covering up Israel's complicity in the World Trade Center attack. And a French couple in Amman, Jordan, had complained about America's "lust for oil." My foray into the Heart of Darkness pushed me over some invisible emotional precipice.

It's easy to play "Bash the Superpower," a sport at which Americans excel. But the recriminations sounded pitiful when spoken by citizens of countries—excuse me, the new One Big Almost-Country—which had posted parks with No Dogs or Chinese Allowed signs, invaded Ethiopia and been hard at work raping and pillaging the Southern Hemisphere long before most people on the planet had ever seen an American GI. Lest we forget: When Washington was still a sleepy swamp, Britain had already seized half the planet, leaving the French and Spanish to divide up most of the rest. The lesser imperialists, like Germany and Holland, had been stuck with the leftovers in Africa and the Americas. Even the Danes got into the act, absorbing poor Iceland and cutting down most of its trees.

Europeans assume that the New Vulgarians, as many call us, are semiliterate bumpkins with no sense of history. They're probably correct—at least, that's what the American tendency to parrot their lame European prattlings without a historical shudder suggests. But in their wonderful education system, don't the British learn that they mucked up badly in Central Asia and the Middle East, then bowed out of the game, leaving us to clean up their mess? Did all of France's schoolchildren skip their lycée lessons the day their teachers explained the impact of their nation's ever-so-sophisticated and nuanced approach to world affairs, as evidenced by the histories of Haiti, Indochina, Algeria and Gabon? And should Americans really revamp our schools and train young people to emulate the civilized behavior of the masters who brought us Stalin, Hitler, Franco, Mussolini and the Spanish Inquisition?

If Europeans want to compete at a round of "Who's committed the greatest number of evil acts on the planet?" America isn't even in contention.

Oh, dear, am I falling into stridency, surely another nasty American habit? Forgive me, but my mother pounded into my head the

old adage about the pot calling the kettle black, an aphorism I suspect French youngsters don't learn. They should, because they need
to know that it's bad form to accuse others of committing sins of
which you yourself are egregiously guilty—say, accusing America
of plotting war against Iraq in pursuit of its oil when your largest
oil company, Total ElfFina, owns the development rights to Iraq's
richest oil field.

I was out of control, as you might already have gathered, and
needed escape, so we flew to Siem Reap to bask in the wonders of
the ancient temples of Angkor Wat. Hot and tired after more than
two months on the road, barely recovered from our two-day expedition up the Mekong and the suffocating dust and humidity of
Phnom Pehn, we took refuge in the magical temple of Ta Prohm,
the only monument in the vast Angkor complex that has been left
to the jungle. The roots of gigantic banyan, kapok and fig trees had
worked their way between the buildings' stones, entwining themselves in the walls and the floors, spreading a canopy of branches
and leaves over the structures. The air was heavy. Only the screeching of monkeys and birds broke the stillness. Crawling through
fallen doorways and lingering in hidden courtyards, we relaxed in
the almost comforting reminder of the impermanence of the
human hand.

Suddenly, we heard the tramping of feet and the discordant
sound of voices—a mumbling Brit, the imperious tones of a couple
of French—expressing relief at the absence of Americans. It was the
by-then familiar conversation about the American freak show and
the European salon: They're prudes; we're enlightened. They're
racists; we're tolerant. They're violent; we're the only moderating
force left in the world.

We didn't emerge from our hiding place, and I bit my tongue—
not metaphorically but literally—to stifle some retort about how
well our "violent nature" served them the last two times they were
unable to disentangle themselves from collective suicide.

They moved on quickly to September 11, and a new voice,
female, German or Dutch, I supposed, quipped with what I imagined was a condescending sneer, "They're making such a big deal
out of it, but not that many people died. They're just brats who

don't understand suffering, and now they're taking advantage of 9/11 to make themselves even more of the superpower."

Clearly, the early sympathy summed up in *Le Monde*'s exuberant WE ARE ALL AMERICANS headline had faded. Stoked by pride in the unification of Europe—an achievement the American states managed more than two centuries earlier, I might add—they were reverting to a full measure of envy, ethnocentrism and postcolonial guilt.

"They have no respect for other people, especially for people who aren't white," one of the French said. "They're not sensitive to others' cultures."

I should have laughed, both because it was the French, who are hardly known for their tolerance or lack of racism, and because of the delicious bathos of the setting. In 1924, one of France's proudest sons, writer André Malraux, had been caught trying to smuggle out of Cambodia 1,800 pounds of sculpted stones he'd prised off Angkor's Banteay Srei Temple. Malraux had gone on to become his country's minister of culture. Paris's Musée Guimet had become Europe's premier museum of Asian art because good Frenchmen like Louis Delaporte had pilfered giant stone heads, statues and carvings from places like Angkor.

The Germans and British had proven themselves equally adept at demonstrating their sensitivity to other cultures by turning international thievery into high culture, a lesson you couldn't miss in Indochina, Asia or the Middle East. When World War I broke out, German archaeologists working in Iraq fled back to Berlin, taking along Babylon's Ishtar Gate and 118 of the 120 friezes that lined the city's Procession Street—and the sensitive leaders of the Pergamon Museum in Berlin aren't even offering to give them back. When Iraq fell under the "protection" of the British, two more shiploads of artifacts disappeared, only to reappear in the British Museum, that great repository of everyone else's culture.

And in 538 B.C.E., more than two millennia before the French Declaration of the Rights of Men and Citizens, Cyrus the Great, founder of Persia's Achaemenid dynasty, proclaimed the planet's first charter of human rights, guaranteeing his people freedom of religion, decent housing, swift punishment of all oppressors and protection of their basic liberties. The nine-inch-long clay barrel on which the cuneiform was inscribed, perhaps the proudest relic of Iranian culture, cannot be seen by Iranians today, unless they fly to a country dedicated to respect for other people—to London, where they have to queue up to see their national treasure at the British Museum.

We hadn't set foot in a theocracy since late January, when we traveled in Iran, nor had we intended to repeat the experience on our way home. So when our plane from Cambodia, via Thailand, landed at the Yangon Airport, I was more than a little disconcerted to find myself in a fundamentalist state.

Myanmar wasn't billed as a religious dictatorship, the government preferring to advertise itself as the dictatorship of the "Peoples' Desires." The concept of theocracy, after all, is as antithetical to Buddhism, the state religion, as is the concept of God.

But the trouble with Myanmar, which most people still persist in thinking of as Burma, to the chagrin of the ruling generals, was that you couldn't spend three days there without falling into some hackneyed rhetorical device like "if it walks like a dog," or "a rose by any other name." And the dog not only walked, but it barked, chewed up slippers and jumped up for treats like a theocracy. So, with my apologies to those who actually understand Buddhism, let me take you on a tour of the world's only fundamentalist Buddhist state.

First, the basics: Buddhism is a theism parading as a nontheism, at least as practiced in Myanmar. Buddha was a teacher, not God. But in Burma, there were 20 million statues of Buddha—reclining Buddha, standing Buddha, sitting Buddha, meditating Buddha, giving Buddha, alabaster Buddha, bronze Buddha, marble Buddha, chalcedony Buddha, bone Buddha, and curvaceous, seductive Buddha with a blinking halo—to which people prayed, genuflected and made offerings, which wasn't bad for a simple teacher who left behind no writings.

All this devotion looked a lot like religion to me, so one rainy afternoon I asked the owner of our hotel in Kyaing Tong to help me understand what seemed like an anomaly.

Befuddled by my confusion, he said only, "Buddhism isn't a religion, it's a philosophy."

That threw me for a loop, since I've never heard of a Cartesian building a hundred-foot-long likenesses of a Reclining Descartes, or a Kantian lusting after a salary increase heaping coins at the altar of old Immanuel. It's tough to imagine an Aristotelian riding a bus all day to worship five of Aristotle's hairs, or a Marxist donating old jewelry so that Karl's image could be enshrined in a structure covered with gold leaf.

So, sue me. I just didn't get it.

The goal of all of this nonworship was to achieve nirvana, a state of noncorporeal being, which looked really enticing when you believed, as did Buddha, that infelicity, impermanence and insubstantiality are the inevitable and inherent hallmarks of the corporeal world—the old, even the best moments are fleeting and empty. Birth leads to sickness, which gives way to old age, which ends in death. The only way out of that cycle of pain and despair was nondesire, the true road to happiness. And you were helped along on that odyssey, improving your karma, by studying, remaining chaste, loving and abstaining from killing, theft, lying, and intoxication.

Myanmar was surely the living embodiment of Buddha's teachings, a place of infelicity, impermanence and insubstantiality where being born led to sickness, old age and death in short order, the median life expectancy not even reaching sixty. I don't know what the Burmese did in their former lives, which revolved around a pretty predictable series of tribal conflicts until the British, French and Dutch decided that their national honors demanded that they annex large swaths of Asia. But its collective karma must be abysmal.

Myanmar was a time warp, almost a stage set dreamed up by some scion of the first world nostalgic for the "unspoiled." Children really did run after the horse-drawn buggies that carried you around town, and the guys driving their ox carts down the lanes actually did smile and wave, just like in the movies. The maid at the Thazin Garden Hotel in Bagan left jasmine petals on our bedspreads, arranged in the shape of hearts. On Lake Inle, villagers grew tomatoes, eggplants, melons and betel vines on floating gardens atop water hyacinth beds. Fishermen propelled their flat-bottomed boats by standing like ostriches and using their free legs to row, and children paddled themselves home after school.

But the maximum monthly government salary of $15 barely covered the cost of food, and the only Asians with less purchasing power than the Burmese were the Afghans. Almost 100,000 of the 1.3 million children born each year died before celebrating their first birthdays. One-quarter of the population lived below minimal subsistence levels. Yet the number of people going to hospitals was decreasing as the price of medical services and drugs became increasingly out of reach.

Myanmar was generating enough income from its monopolies on teak, jade, rubies and sapphires to feed its people, albeit not with filet mignon. But the government spent 40 percent of that income on an oversize military and an endless number of economically dubious schemes that reinforced the local students' joke that the combined university admission test scores of all four ruling generals wouldn't have been high enough to admit even one of them.

As we traveled from Mongolia into China and then Indochina, the local modes of transportation had moved backward in time. In China, cars were edging out bicycles, while the Vietnamese had to content themselves with motorcycles. Motorized transportation was just beginning to make an appearance in Cambodia, but in Myanmar, people were lucky to have bicycles. On the hour drive between the airport at Heho to Lake Inle, we passed only four vehicles: they were all tiny pickup trucks, one with nine people crammed inside and another thirteen hanging onto the top.

Even the oversize NGO or UN Toyota Land Cruisers were absent, the generals having decided that Myanmar didn't need foreign aid. "Myanmar is not a beggar state," the politically correct declared proudly. "It's Buddhist tradition."

For forty years, Burma had been governed by a military "Revolutionary Council" that replaced postindependence fictive democracy with nonfictive despotism. Thanks to their homegrown political strategy, the "Socialist Path to Nirbanna," by the time we arrived, the people of Myanmar had been living under one of the world's most repressive regimes for forty years. Citizens were banned from discussing politics with foreigners, and government workers were required to sign a pledge not to talk politics with one another. When the 400,000 men in uniform proved insufficient, Myanmar's gang of four—the generals who ran the State Law and Order Restoration Council, popularly known as SLORC, which later performed a facelift by firing two members and renaming itself the State Peace and Development Council—forcibly conscripted ordinary folk to serve as porters and human mine sweepers, or to work as "volunteers" on beautification and historical restoration projects. Foreign music was outlawed, foreign dress severely discouraged, and an army of paid informants ferreted out "minions of colonialism."

Myanmar was a lot like an American public high school, the rules rigid, the hectoring admonishments ubiquitous. The elevator at Yuzana Garden Hotel in Yangon didn't stop at the second floor, the sign explaining that curiosity reading, "Climbing Steps Is Good Exercise." Immense billboards decorated every town advertising "The Peoples Desires," which included, in Burmese and English, the wish to "Oppose those relying on external elements, acting as stooges, holding negative views," and to "Crush all internal and external destructive elements as the common enemy."

Not all the Burmese were starving—not in the world's largest producer of heroin. In the Golden Triangle, where Laos, Thailand and Myanmar meet in a remote monsoon forest, the parking lot at the Seik Tie Kaw Restaurant was jammed with four-wheel-drive vehicles and pickup trucks, one cherry red with a shiny chrome roll bar. Men lazed at the tables, sharing bottles of Johnny Walker Red on the rocks, the only ice we saw outside four-star hotels in Yangon.

High in the hills, we trekked up the muddy slopes of the forest for hours to isolated villages of hill tribesmen, animist people still living in traditional multiple-family longhouses that teeter on stilts. At first glance, the villagers seemed untouched by modernity, eating only what they grew, wearing traditional clothing they wove and dyed by hand. Then a man strolled up from the river to his hut lugging a massive chain saw, his reward for growing acres of poppies for drug dealers who maintained portable labs in the jungle.

"Who would grow potatoes, which sell for five cents a kilo, when they can grow poppies for heroin, which goes for half a million dollars?" joked our host at the Private Hotel, the former police chief of the region. His affect was jocular, which seemed a strange response from a man who'd spent most of his life in uniform.

"It's payback, you know," he finally added, joking, as did most Burmese, that the generals skimmed 20 percent of the heroin take. "Think about the Opium Wars. The British introduced drugs into this region, now we're returning the favor."

Despite the reign of terror and the grinding poverty, the Burmese accepted their fate with graceful resignation. After all, according to Theravada Buddhism, the main school of Buddhism in Myanmar, individuals are responsible for their own destinies. And the government lavished gifts on the nation's 350,000 monks, the only potential counterforce to its 400,000-member military, to reinforce the message uppermost in the popular mind: Don't blame the government if you don't have enough to eat or your husband dies of malaria. It's your karma. Blackouts regularly envelop the country, but, hey, you need to rid yourself of desire.

The Burmese had an odd reaction to outsiders. We were a novelty, to say the least, since the generals had declared most foreigners personae non grata during the 1960s. When the restrictions were lifted in the late 1980s, Western tourists shied away from Myanmar in deference to a travel boycott advocated by leaders of the opposition. For Burmese, isolated in a pariah nation, the presence of strangers, then, represented the possibility of normalcy in a land where normalcy was a phantasm for all but those old enough to evoke the distant memory of "before," or the few who could afford the $1,000 cost of a passport.

But as happy as Burmese were to see us, they were also miffed, or at least as miffed as Burmese, who have a strong cultural bias against displays of anger, ever get. Their annoyance, however, was strangely out of sync with the international zeitgeist, or at least what most Americans and Europeans believed that sentiment to be.

"When Clinton imposed economic sanctions on Myanmar, he did nothing to hurt the regime," one man explained. Those sanctions, imposed in May 1997, included a ban on U.S. investment in oil and natural gas development and prohibited U.S. citizens from entering into contracts for the development of Burmese economic resources. "The generals said, 'See if we care, we can do without you.' And the generals are doing fine without America. But the rest of us. . . ."

Another man, educated and a firm supporter of Nobel laureate Aung San Suu Kyi, the symbol of Burmese resistance, was even blunter: "Why won't the U.S. do for Myanmar what it did for Afghanistan?"

I didn't have the heart to answer, "Buddhists make lousy terrorists."

The two-day flight back from Burma left us reeling with exhaustion, and after all the miles we had traveled and all the mishaps we had endured, there was nothing—nothing—we wanted more than the reassurance and the comfort of home. But we knew that once we got there, we were unlikely to emerge for a while. So we summoned the strength to take a detour from our three-hour drive to the all-night Stop & Shop. It had never made much of an impression on me before. It had simply been normal. But as I wandered the aisles, I could no longer take the plenty, or the convenience, for granted. It made me dizzy, in the best of ways. It even filled me with a strange kind of joy. Strange, because its prompts were so mundane: two dozen kinds of cereal, a plump Perdue chicken, a ripe California avocado.

It was still dark when we turned onto our 1.5-mile driveway on August 25. But just around the first bend, I spotted a sign tacked to a tree, faintly visible in the predawn light: PROUD TO BE AN AMERICAN. Too many of my instincts were honed at demonstrations during the 1960s for me to feel at ease with such overt expressions of patriotism, so I flinched. And yet, to my surprise, that impulse only lasted an instant. My stronger reaction was a characteristically editorial one. I wanted to cross out the word PROUD and replace it with GRATEFUL. Which I was, and not only for the bounty that lined the aisles of the supermarket. For the order and variety of it all, the triumph of choice that it represented, the constancy it promised. It was gluttony, yes, and it was indulgence.

But there was not a person I had met in the places I had traveled who would not have embraced it, and relished it, and been elated to call it their own.

We unpacked—bit by bit, bag by bag, over hours made sluggish by our spent energies. A dozen chores loomed like an intricate, infinite nightmare. We had to reconnect our utilities. We had to reregister our cars. We had to have propane delivered, plumbing checked. And yet we sped through these tasks in glorious efficiency, thanks to modern computers and clerks trained to service and not snarls.

For most of the first week back, we did not leave our mountaintop, oddly reluctant to reengage. For an hour or two each day, we let CNN break the silence, comforted that we had an actual source of information after a year when even facts as basic as the number of casualties in Aksy had been unobtainable.

But the reports and dissections of tsunamis of anti-Americanism sweeping across the globe and of the alarming rise of fundamentalism in Central Asia left me feeling like Alice in CNN's Wonderland. After a year sliding down rabbit holes into Baghdad and Tehran and trying to make sense out of my students' jabberwocky, I'd looked forward to home, where everything and everyone, myself included, would be the right size and operate according to the right cadences and the right reasoning.

But the world I fell back into seemed as strange as Bishkek, Tehran or the Queen's croquet court. It was not only a wondrous place of avocadoes and juicy chickens, but also one in which a dozen blowhards from the left, right and center limned reality in the rhymes of Dodo and the Duck, speaking to their own logic, taking figurative language literally, literal language figuratively and never waiting their turn. Everything I heard on television sounded entirely, utterly, crystal-clear-edly lucid. But it made no sense whatsoever outside of that parallel universe where the balls were live hedgehogs, the mallets pink flamingoes, and the pundits spent more time in the makeup room than in the places they yammered about.

In the other world, the one where I'd spent the year, our friends weren't always noble, our enemies weren't ineluctably venal, and the people—those near-mythic "ordinary people"—rarely followed the script we believed they were writing. Muslims from Central Asia and the Middle East were drawn to Al-Qaeda's jihad the same way American socialists and fellow travelers had been drawn to fight Franco's fascists in Spain in the 1930s—with passion, but in minuscule numbers. For them, the struggle was a contest of cosmic proportions, a battle both against Valentine's Day and Nike and against western concepts like tolerance and human rights that threatened the foundations of their universe.

Anti-Western sentiment was a palpable presence. But it wasn't set off so much by ire at the multinational corporations that dictated the earth's orbit as by zealots defending old-world traditions like virginity and veiling and the unease of the masses about their place in the future. The masses had already donned jeans, at least in the privacy of their own homes, and they were watching MTV. But still caught between those often-fanatical forces of the past and the relentless tug of modernity, they were plagued by the sense that, in the process, their tribal ties were dissolving, that they were losing their identities. If progress didn't fulfill its promise, they could turn on a dime. But they were aching not to be forced to.

From my new vantage point, with one foot in New York and the other still dragging behind me somewhere along the road from Bishkek, every time Americans tried to make things better—either by rattling a saber or by espousing what we thought of as enlightened, progressive policies—we wound up making them worse. Even if the United States became everyone's paragon of virtue—which would be a twisted tango given the vagueness and endless mutations of virtue's definition—I wondered how much really would change. Could I count on the Russian press or Iranian television to share that good news with people with no means to change the channel?

As late summer mellowed into the crimson of early fall and America grappled with the painful anniversary of September 11, Dennis and I relearned daily that Americans didn't know enough, didn't understand enough. But we were also reminded that non-Americans didn't either. If they're not whom we think they are, neither are we who they perceive, caricatures of rich cocaine-snorters, greedy materialists and satanic infidels all culled from characters flitting across movie screens, out of meticulously censored press reports and the blare of the haters.

The distance between us and them was a chasm of nothing so simple as geography or uneven levels of wealth. It was a broader failure of understanding—forged by suspicion, misinformation and divergent historical paths—and it was not easy to bridge, although the instinct to try was keen on both sides. My thoughts kept returning to the afternoon I spent crowded into the stands of the old soccer stadium in Sapa, Vietnam, among Hmong and Dzao villagers. As exuberant teenagers kicked balls on the dusty field, the local women—wearing an amazing assortment of tribal hats, some festooned with old silver coins, others with long fringe, still more decorated with elaborate embroidery—studied me openly, occasionally testing their English with a pointed question about my home and my family. I returned their frank glances, their curiosity, frustrated at our clumsy attempts at communication, realizing that we were all trying to glean enough information, a few facts, a couple of details, to . . . imagine.

It was an impossible feat, of course. Those tiny women, raised in villages high in the Hoang Lien Son Mountains, could no more imagine my $8,000 property tax bill or the smell of the New York subway at rush hour on a steamy summer afternoon than I could conjure up the reality of hauling buckets of water up a mile of hill, or being gored by a water buffalo three hours by foot from the nearest physician—no matter how fervently we believed that we could make that leap.

Imagining the lives of others is an essential human instinct. It is an act of empathy, a gesture of faith in a common bond. It is also a kind of travel, an attempt to move outside the parameters of our own narrow universe. But it almost always fails. Once we pick up and go—once we cross the borders, physical and intellectual, political and emotional, that divide countries and continents—we come to realize that we're not merely imagining. We're projecting. And if we're honest with ourselves about that, we at last see the truths, or at least the puzzles and muddles in which they are buried.

Acknowledgments

When eating bamboo sprouts,
remember the man who planted them.
—CHINESE PROVERB

For much of my life, I've written about government programs that don't work, so it is with a particular sense of relief—even joy—that I express my immense gratitude to one that does. For more than a half century, Fulbright programs have been turning J. William Fulbright's vision of building peace and understanding through the exchange of individuals and ideas into reality. Thousands of us have enjoyed the extraordinary opportunity, to immense personal and international benefit. My heartfelt thanks, then, go to the Fulbright Commission and to the Center for the International Exchange of Scholars, especially to Dr. Andy Riess, for giving me the opportunity that became this book.

Most Americans assume that our diplomats spend their days sipping tea with the powerful and their nights attending swanky soirées. Perhaps they do in places other than Central Asia, but in Bishkek, they work hard on behalf of the citizens of America, and they labored mightily to make my stay in Bishkek both pleasant and productive. My thanks go to them all, most especially to Conrad Turner and his dedicated staff.

My trips to Uzbekistan, Turkmenistan and Mongolia were sponsored by the U.S. embassies in those countries, and I am indebted to Mark Asquino, Michele Ullrich, and Theresa Markiw, the public affairs officers in Tashkent, Ashgabat and Ulan Baator and their exceptional staffs, as well as to Ravdan Oyuntsetseg, Valia Papoutsaki and the Mongolian Press Institute for showing me, at long last, Ulan Baator.

The students of the international journalism program at the Kyrgyz-Russo-Slavic University were my finest guides through this moment in Central Asian history, and I am deeply grateful to them for putting up with my incessant questions, my weekly hectoring and my attempts at playing excavator.

Part of me will always consider Bishkek to be a second home, in large measure because of the generosity of friendship offered me by Mamasadyk Bagyshov, Victoria Badiukova, Vitaly Gergert, Gulnara Bekbolinova, Azamat Aitoktor uulu, Konstantin Sudakov, Ahoura Afshar, Regina Gatina, Samarbek Ashym uulu, Meghan Simpson, and Norma Jo Baker. My thanks to them all.

Marina, Vladimir, Nikolai and Artyom Zhdanov showed us Russia as I only hope I can one day show them America. я глубоко благодарна вам.

My journey into Afghanistan was a gift given me by Laurie Abraham and Roberta Myers of *Elle*, one of the many gifts they have given me over the years.

Along the way, scores of clerks and taxicab drivers, pedicab owners, guides, waiters, conductresses and passersby dropped a word or a look or a full conversation that grew into this book. They'll never know how much they contributed to a stranger's understanding. But I won't forget their kindness.

My warmest thanks to the team at HarperCollins, in particular to Susan Weinberg, Cathy Hemming and Alison Callahan, for nurturing this project, and for so much more. Lisa Bankoff has put up with me for eight books, which earns her a special medal of recognition, and Patrick Pryce has made both of our lives more bearable.

Writers rise and fall on the patience of their friends, and if I've met the challenge, it is in large measure because mine have been so consistently willing to put up with my endless storytelling, my angst and my pleas for help. This book bears the indelible stamp of four extraordinary men whom I must single out in gratitude. The inimitable Robert Jones turned me into a writer of books by refusing to allow me to complain. For fifteen years, Frank Bruni has never failed me when my own words did. Nick Busse wrote me not only the first editorial letter I ever received, but one filled with candor and insight. And Stephen Ray changed the shape of the world.

Finally, most latter-day Lucys aren't fortunate enough to find their Rickys, their partners in crime, and in passion. I'm one of the lucky ones.